T0219654

.

Automotive Technician Training: Practical Worksheets Level 2

Automotive practical worksheets for students at Level 2

This Level 2 student worksheets book contains tasks that help you develop practical skills and prepare you for assessment. The tasks also reinforce the automotive theory that you will learn online and in the classroom. Each worksheet covers individual topics in a step-by-step manner, detailing how to carry out all the most important tasks contained within the syllabus. Alongside each of these worksheets is a job card that can be filled in and used as evidence towards your qualification.

▶ Endorsed by the Institute of the Motor Industry for all their Level 2 automotive courses.
▶ Step-by-step guides to the practical tasks required for all Level 2 qualifications.
▶ Job sheets for students to complete and feedback sheets for assessors to complete.

Tom Denton is the leading UK automotive author with a teaching career spanning lecturer to head of automotive engineering in a large college. His range of automotive textbooks published since 1995 are bestsellers and led to his authoring of the Automotive Technician Training multimedia system that is in common use in the UK, USA and several other countries. Tom now works as the eLearning Development Manager for the Institute of the Motor Industry (IMI).

Automotive Technician Training

Practical Worksheets Level 2

Tom Denton

Routledge
Taylor & Francis Group

LONDON AND NEW YORK

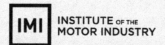

INSTITUTE OF THE
MOTOR INDUSTRY

First published 2015
by Routledge
2 Park Square, Milton Park, Abingdon, Oxon OX14 4RN

and by Routledge
711 Third Avenue, New York, NY 10017

Routledge is an imprint of the Taylor & Francis Group, an informa business

© 2015 Tom Denton

The right of Tom Denton to be identified as author of this work has been asserted by him in accordance with sections 77 and 78 of the Copyright, Designs and Patents Act 1988.

All rights reserved. The purchase of this copyrighted material confers the right on the purchasing institution to photocopy pages which bear the copyright line at the bottom of the page. No other parts of this book may be reprinted or reproduced or utilised in any form or by any electronic, mechanical, or other means, now known or hereafter invented, including photocopying and recording, or in any information storage or retrieval system, without permission in writing from the publishers.

Trademark notice: Product or corporate names may be trademarks or registered trademarks, and are used only for identification and explanation without intent to infringe.

British Library Cataloguing-in-Publication Data
A catalogue record for this book is available from the British Library

Library of Congress Cataloging in Publication Data
A catalog record for this book has been requested

ISBN: 978-1-138-85237-2 (pbk)
ISBN: 978-1-315-72351-8 (ebk)

Typeset in Univers by
Servis Filmsetting Ltd, Stockport, Cheshire

Practical Worksheets – Level 2

Introduction

The purpose of this worksheets book is to provide a range of practical activities that will enable you to develop your abilities as a technician. The tasks are aligned with recognized vocational qualifications. However, there are far more tasks within this workbook than are required by the awarding body for the achievement of a Vocationally Recognized Qualification – because the more you practise, the more skills you will develop.

The worksheets are presented as three separate books at Level 1, Level 2 and Level 3 to follow the recognized qualifications. Within each level there are tasks for the major automotive areas: Engines, Chassis, Transmission and Electrical. The tasks range from component identification to removal and refit at Level 1 and 2, and diagnosis of complex system faults at Level 3.

A blank job card and assessor report are provided with each worksheet. This should be copied and then filled in alongside the task you are completing, including all relevant details regarding the vehicle, the fault and the rectification procedure where appropriate. You should write down a description of the work that you did to complete the task including any technical data that you sourced, any difficulties that you encountered and how you overcame them. If you had any health and safety issues to address, i.e. disposal of waste materials or clearing up spillages, this will help demonstrate your competence. By completing job cards thoroughly at this stage of your career as a technician, you will be well prepared for the time when you are required to complete job cards in the workplace. This can be very important, for example, if a warranty job card is not accurate then the manufacturer will not pay for the claim. An example of a completed job card is shown on page 7.

For teacher/lecturers, this work book more than covers the requirements for Vocational Qualifications. Using the following tracking document you can note progress and also cross-reference the highlighted worksheets that directly relate to the awarding body required practical tasks.

Tracking

Engines (p. 8)	11	22	33	43
1	12	23	34	44
2	13	24	35	45
3	14	25	36	46
4	15	26	37	47
5	16	27	38	48
6	17	28	39	49
7	18	29	Chassis (p. 86)	50
8	19	30	40	51
9	20	31	41	52
10	21	32	42	53

54	59	65	70	76
55	60	**Electrical** (p. 138)	71	77
56	61	66	72	78
57	62	67	73	
Transmission (p. 122)	63	68	74	
58	64	69	75	

Important notes about practical work

Safety

Working on vehicles is perfectly safe as long as you follow proper procedures. For all of the worksheets in this book you must therefore:

Comply with personal and environmental safety practices associated with clothing; eye protection; hand tools; power equipment; proper ventilation; and the handling, storage, and disposal of chemicals/materials in accordance with all appropriate safety and environmental regulations.

There are some specific recommendations below but you should also refer to the other textbooks or online resources for additional information.

Personal protective equipment (PPE), such as safety clothing, is very important to protect yourself. Some people think it clever or tough not to use protection. They are very sad and will die or be injured

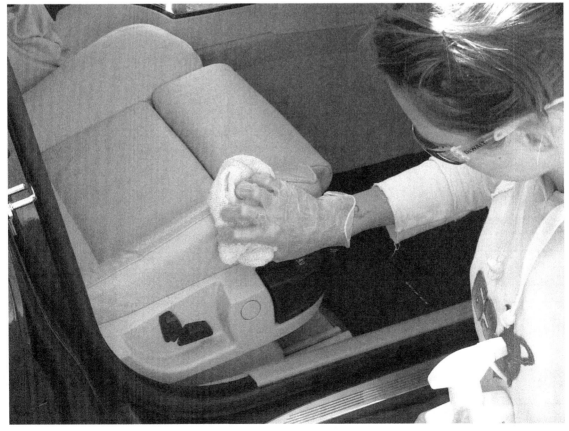

Eye protection and gloves in use

long before you! Some things are obvious, such as when holding a hot or sharp exhaust you would likely be burnt or cut! Other things such as breathing in brake dust, or working in a noisy area, do not produce immediately noticeable effects but could affect you later in life.

Fortunately the risks to workers are now quite well understood and we can protect ourselves before it is too late. In the following table, I have listed a number of items classed as PPE (personal protective equipment) together with suggested uses. You will see that the use of most items is plain common sense.

Equipment	Notes	Suggested or examples where used
Ear defenders	Must meet appropriate standards	When working in noisy areas or if using an air chisel
Face mask	For individual personal use only	Dusty conditions. When cleaning brakes or if preparing bodywork
High visibility clothing	Fluorescent colours such as yellow or orange	Working in traffic such as when on a breakdown
Leather apron	Should be replaced if it is holed or worn thin	When welding or working with very hot items
Leather gloves	Should be replaced when they become holed or worn thin	When welding or working with very hot items and also if handling sharp metalwork
Life jacket	Must meet current standards	Use when attending vehicle breakdowns on ferries!
Overalls	Should be kept clean and be flame proof if used for welding	These should be worn at all times to protect your clothes and skin. If you get too hot just wear shorts and a T-shirt underneath
Rubber or plastic apron	Replace if holed	Use if you do a lot of work with battery acid or with strong solvents
Rubber or plastic gloves	Replace if holed	Gloves must always be used when using degreasing equipment
Safety shoes or boots	Strong toe caps are recommended	Working in any workshop with heavy equipment
Safety goggles	Keep the lenses clean and prevent scratches	Always use goggles when grinding or when any risk of eye contamination. Cheap plastic goggles are much easier to come by than new eyes
Safety helmet	Must be to current standards	Under vehicle work in some cases
Welding goggles or welding mask	Check the goggles are suitable for the type of welding. Gas welding goggles are NOT good enough when arc welding	You should wear welding goggles or use a mask even if you are only assisting by holding something

Also, as well as your own protection you should always use a protection kit for the vehicle: floor mats, wing covers and seat covers for example.

Tools and equipment

To carry out any work you will need a standard toolkit and in some cases additional 'special' tools will be required. Make sure you have access to all necessary equipment before starting work. A few examples are mentioned below but you should also refer to the other textbooks or online resources for additional information.

Using hand tools is something you will learn by experience, but an important first step is to understand the purpose of the common types. This section therefore starts by listing some of the more popular tools, with examples of their use, and ends with some general advice and instructions.

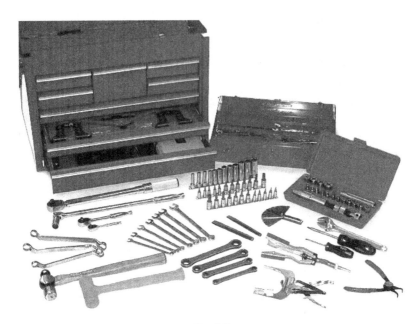

Toolkit

Practise until you understand the use and purpose of the following tools when working on vehicles.

Hand tool	Example uses and/or notes
Adjustable spanner (wrench)	An ideal stand by tool and useful for holding one end of a nut and bolt.
Open-ended spanner	Use for nuts and bolts where access is limited or a ring spanner can't be used.
Ring spanner	The best tool for holding hexagon bolts or nuts. If fitted correctly it will not slip and damage both you and the bolt head.
Torque wrench	Essential for correct tightening of fixings. The wrench can be set in most cases to 'click' when the required torque has been reached. Many fitters think it is clever not to use a torque wrench. Good technicians realize the benefits.
Socket wrench	Often contain a ratchet to make operation far easier.
Hexagon socket spanner	Sockets are ideal for many jobs where a spanner can't be used. In many cases a socket is quicker and easier than a spanner. Extensions and swivel joints are also available to help reach that awkward bolt.
Air wrench	These are often referred to as wheel guns. Air-driven tools are great for speeding up your work but it is easy to damage components because an air wrench is very powerful. Only special, extra strong, high-quality sockets should be used.
Blade (engineer's) screwdriver	Simple common screw heads. Use the correct size!
Pozidrive, Philips and cross-head screwdrivers	Better grip is possible particularly with the Pozidrive but learn not to confuse the two very similar types. The wrong type will slip and damage will occur.
Torx®	Similar to a hexagon tool like an Allen key but with further flutes cut in the side. It can transmit good torque.
Special purpose wrenches	Many different types are available. As an example mole grips are very useful tools as they hold like pliers but can lock in position.
Pliers	These are used for gripping and pulling or bending. They are available in a wide variety of sizes. These range from snipe nose, for electrical work, to engineers pliers for larger jobs such as fitting split pins.
Levers	Used to apply a very large force to a small area. If you remember this you will realize how, if incorrectly applied, it is easy to damage a component.
Hammer	Anybody can hit something with a hammer, but exactly how hard and where is a great skill to learn!

General advice and instructions for the use of hand tools (taken from information provided by Snap-on):

▶ Only use a tool for its intended purpose
▶ Always use the correct size tool for the job you are doing
▶ Pull a spanner or wrench rather than pushing whenever possible
▶ Do not use a file or similar without a handle
▶ Keep all tools clean and replace them in a suitable box or cabinet
▶ Do not use a screwdriver as a pry bar
▶ Look after your tools and they will look after you!

Information

Before starting work you should always make sure you have the correct information to hand. This can be in the form of a workshop manual or a computer-based source.

The worksheets in this book are a general guide so make sure the correct information, procedures and data for the particular vehicle you are working on are available before you start work.

 Technical data

 Timing belts

 Timing chains

 Timing gears

 Service indicator reset procedures

 Key programming

Manufacturers' service schedules

 Service illustrations

 Repair times

Wheel alignment

 Diagnostic trouble codes

 Tyre sizes and pressures

 Known fixes and bulletins

 Engine management component testing

 Engine management pin data

 Engine management trouble shooter

 Airbags

 Anti-lock brake systems

 Air conditioning

 Electrical component locations

 Wiring diagrams

 Guided diagnostics

 Tyre pressure monitoring system

 Electric parking brake

 Battery disconnection and reconnection procedures

Autodata online information

Job card: example

Technician/Learner name & date	Make and model	VIN no.	Reg. no.	Job/task no.
John Doe	Ford Mondeo	1M8GDM9A_KP042788	ABC 123	100

Customer's instructions / Vehicle fault	Mileage	
	67834	

Customer's instructions / Vehicle fault · **Mileage** 67834

Carry out minor service – change oil and filter.
Inspect brakes.
Check for rattle underneath when accelerating.

Work carried out and recommendations (include PPE & special precautions taken)

PPE worn – boot, gloves and overall, VPE – wing covers, floor mats and steering wheel cover. Followed service checklist, full under bonnet check of belts, for oil, fuel and coolant leaks. Drained oil and replaced filter, torque new filter to 15Nm as per manufacturer's instructions, filled engine with 6 litres of engine oil. Old engine disposed of in oil drum and filter placed in specific filter bin.

Full under vehicle check, hoses, brakes pipes, all steering and suspension components – all ok. Found detached exhaust mounting – this would cause the knock the customer complained of, replaced mounting.

Checked brakes, pads and discs ok 30% wear on pads.

Requires 2 front tyres, have notified customer but they will go to Kwikfit.

During the service a quantity of oil was spilled on the floor, I covered the spillage with granules and left them to soak the oil up. I then cleaned the granules up and disposed of them in the correct bin. Finally I mopped the floor to ensure that it was properly clean.

Parts and labour	Price
3 hours @ 22.50 per hour	£67.50
Oil	£18
Oil Filter	£6.80
Exhaust mounting	£14
Total	£106.30

Data and specifications used (include the actual figures)

Oil filter torque – 15Nm, Wheel nut torque – 160Nm, 6.0 litres of engine oil

Assessor report: example

	Assessment outcome	Passed (tick ✓)
1	The learner worked safely and minimised risks to themselves and others	✓
2	The learner correctly selected and used appropriate technical information	✓
3	The learner correctly selected and used appropriate tools and equipment	✓
4	The learner correctly carried out the task required using suitable methods and testing procedures	✓
5	The learner correctly recorded information and made suitable recommendations	✓

Assessor name (print)	Tick	Written feedback (with reference to assessment criteria) must be given when a learner is referred
PASS: I confirm that the learner's work was to an acceptable standard and met the assessment criteria of the unit	✓	*Candidate worked in a very organised manner.* *Work area was kept clean and tidy throughout, tools returned to toolbox once used and oil and filter disposed of correctly.* *Good communication regarding further work requirements found during the task.* *Assessment criteria met, well done.*
REFER: The work carried out did not achieve the standards specified by the assessment criteria		

Assessor Name (Print)	Assessor PIN/Ref.	Date
Jane Jones	1234	29/02/17

The section below is only to be completed by the learner once the assessor decision has been made and feedback given to learner

I confirm that the work carried out was my own, and that I received feedback from the Assessor	Learner name (Print)	Learner signature	Date
	John Doe	J Doe	29/02/17

Engines

Worksheet 1: Routine vehicle maintenance inspections/service

Procedure

▶ General visual inspection – listen for abnormal noises.

▶ Visual/oil level and condition. Replace oil and filter at specified intervals.

▶ Visual/inspections for oil leakage.

▶ Visual/inspection of exhaust smoke – at idle speed at mid-engine speed (3000 rpm) on overrun during road test.

▶ Engine oil pressure test – check warning light operates (attach a pressure gauge and adapter and tachometer) – pressure at idle speed – stabilized pressure – stabilized pressure engine rpm.

▶ Crankcase ventilation system – check condition of hoses (visual and remove hoses and valves) – orifice to inlet manifold clear – air cleaner condition – control valve condition.

▶ Check coolant level and specific gravity.

▶ Check brake operation and pad/disc condition – record thickness.

▶ Check tyre tread depths.

▶ Torque wheel nuts – include torque figures and calibration date of torque wrench.

▶ Additional items – see manufacturer's schedule.

Job card

Technician/learner name & date	Make and model	VIN no.	Reg. no.	Job/task no.

Customer's instructions/vehicle fault	Mileage	

Work carried out and recommendations (include PPE & special precautions taken)

Parts and labour	**Price**
Total	

Data and specifications used (include the actual figures)

Assessor report

	Assessment outcome	Passed (tick ✓)
1	The learner worked safely and minimised risks to themselves and others	
2	The learner correctly selected and used appropriate technical information	
3	The learner correctly selected and used appropriate tools and equipment	
4	The learner correctly carried out the task required using suitable methods and testing procedures	
5	The learner correctly recorded information and made suitable recommendations	

	Tick	Written feedback (with reference to assessment criteria) must be given when a learner is referred
Pass: I confirm that the learner's work was to an acceptable standard and met the assessment criteria of the unit		
Refer: The work carried out did not achieve the standards specified by the assessment criteria		

Assessor name (print)	Assessor PIN/ref.	Date

Section below only to be completed by the learner once the assessor decision has been made and feedback given

I confirm that the work carried out was my own, and that I received feedback from the Assessor	Learner name (print)	Learner signature	Date

Worksheet 2: Vehicle inspections

Procedure

▶ General visual inspection – listen for abnormal noises.

▶ Check brake operation and pad/disc condition – record thickness.

▶ Check tyre tread depths.

▶ Torque wheel nuts – include torque figures and calibration date of torque wrench.

▶ Additional items – see manufacturer's schedule depending upon type of inspection – pre- and post-work, pre-delivery inspection (PDI), pre-purchase inspection (PPI), pre MOT, visual health check (VHC) or post repair.

Job card

Technician/learner name & date	Make and model	VIN no.	Reg. no.	Job/task no.

Customer's instructions/vehicle fault	Mileage	

Work carried out and recommendations (include PPE & special precautions taken)

	Price
Parts and labour	
Total	

Data and specifications used (include the actual figures)

Assessor report

	Assessment outcome	Passed (tick ✓)
1	The learner worked safely and minimised risks to themselves and others	
2	The learner correctly selected and used appropriate technical information	
3	The learner correctly selected and used appropriate tools and equipment	
4	The learner correctly carried out the task required using suitable methods and testing procedures	
5	The learner correctly recorded information and made suitable recommendations	

	Tick	Written feedback (with reference to assessment criteria) must be given when a learner is referred
Pass: I confirm that the learner's work was to an acceptable standard and met the assessment criteria of the unit		
Refer: The work carried out did not achieve the standards specified by the assessment criteria		

Assessor name (print)	Assessor PIN/ref.	Date

Section below only to be completed by the learner once the assessor decision has been made and feedback given

	Learner name (print)	Learner signature	Date
I confirm that the work carried out was my own, and that I received feedback from the Assessor			

Worksheet 3: Adjust valve clearances (OHV)

Procedure

▶ Run engine and listen for abnormal noises from top of engine. Check for misfire/cylinder balance. Disconnect battery earth/ground cable. Remove rocker cover.

▶ Measure with feeler gauges all valve clearances at the 'back-of-the-cam' (heel) position. Check manufacturer's recommended procedure for positioning the engine, e.g. rule of nine or valves rocking method.

▶ Adjust each valve using a special tool or wrench and screwdriver as appropriate. Note that inlet and exhaust settings may vary.

▶ Refit all and run engine to check for noise or misfire.

Job card

Technician/learner name & date	Make and model	VIN no.	Reg. no.	Job/task no.

Customer's instructions/vehicle fault	Mileage	

Work carried out and recommendations (include PPE & special precautions taken)

Parts and labour	Price
Total	

Data and specifications used (include the actual figures)

Assessor report

	Assessment outcome	Passed (tick ✓)
1	The learner worked safely and minimised risks to themselves and others	
2	The learner correctly selected and used appropriate technical information	
3	The learner correctly selected and used appropriate tools and equipment	
4	The learner correctly carried out the task required using suitable methods and testing procedures	
5	The learner correctly recorded information and made suitable recommendations	

	Tick	Written feedback (with reference to assessment criteria) must be given when a learner is referred
Pass: I confirm that the learner's work was to an acceptable standard and met the assessment criteria of the unit		
Refer: The work carried out did not achieve the standards specified by the assessment criteria		

Assessor name (print)	Assessor PIN/ref.	Date

Section below only to be completed by the learner once the assessor decision has been made and feedback given			
I confirm that the work carried out was my own, and that I received feedback from the Assessor	Learner name (print)	Learner signature	Date

Worksheet 4: Inspect camshaft lobes, journals and bearings, auxiliary shafts, bearings and drive

Procedure

▶ Strip to access the camshaft. Check the end float before removing.

▶ Remove camshaft and visually inspect for condition of bearing journals and cam lobes. Inspect retaining/thrust plate for wear.

▶ Inspect bearings in the engine block or head. Look for wear, scoring, pitting, or other deterioration.

▶ Lay the camshaft in oiled paper supports in vee blocks on a surface table. Assemble a dial test indicator (DTI) to a base and zero on centre journal. Rotate shaft and observe variation in needle position to indicate if the shaft is bent.

▶ Measure and record cam lift on all lobes. Compare with manufacturer's data.

▶ Measure cam journals with external micrometer and record results. Measure for ovality. Measure internal dimensions of bearings and record. Compare readings with manufacturer's data for size and wear tolerances.

▶ Reassemble shaft and other components. Run engine and listen for abnormal noises.

▶ The inspection of auxiliary shafts is a similar procedure.

Job card

Technician/learner name & date	Make and model	VIN no.		Reg. no.	Job/task no.

Customer's instructions/vehicle fault	Mileage	

Work carried out and recommendations (include PPE & special precautions taken)

	Price
Parts and labour	
Total	

Data and specifications used (include the actual figures)

Assessor report

	Assessment outcome	Passed (tick ✓)
1	The learner worked safely and minimised risks to themselves and others	
2	The learner correctly selected and used appropriate technical information	
3	The learner correctly selected and used appropriate tools and equipment	
4	The learner correctly carried out the task required using suitable methods and testing procedures	
5	The learner correctly recorded information and made suitable recommendations	

	Tick	Written feedback (with reference to assessment criteria) must be given when a learner is referred
Pass: I confirm that the learner's work was to an acceptable standard and met the assessment criteria of the unit		
Refer: The work carried out did not achieve the standards specified by the assessment criteria		

Assessor name (print)	Assessor PIN/ref.	Date

Section below only to be completed by the learner once the assessor decision has been made and feedback given

I confirm that the work carried out was my own, and that I received feedback from the Assessor	Learner name (print)	Learner signature	Date

Worksheet 5: Remove and reinstall engine (RWD) without transmission

Procedure

▶ Obtain crane, sling and lifting eyes with a SWL in excess of engine weight. Disconnect battery earth/ground cable. Fit lifting eyes to engine as per manufacturer's instructions. Label and disconnect all electrical cables to engine including engine ground cable. Label and disconnect all vacuum pipes to engine. Remove air intake ducting and air cleaner, blank off open pipes.

▶ Depressurize, label and disconnect fuel lines to engine, blank off pipes. Label and disconnect throttle/choke and other control cables. Disconnect exhaust down pipe at manifold flange. Remove, if necessary, the starter motor and clutch operating mechanism (cable or hydraulic). Slacken all clutch housing bolts and remove all but two, one either side. Slacken and remove, if safe to do so, the engine mounting nuts from the underside of vehicle.

▶ When all underside work is completed, drain the coolant and remove radiator and all hoses. When air conditioning is fitted, do not disconnect refrigerant pipes/hoses before removing refrigerant (specialist task). If possible, remove and support the compressor with hoses attached.

▶ Attach sling to lifting eyes and crane, tension sling. Check all connections to engine have been disconnected. Remove engine mounting nuts/bolts if not already removed. Operate crane to lift engine clear of mountings. Put a jack under the gearbox and lift slightly. Remove remaining clutch housing bolts. Pull engine forward away from the gearbox. Lift slowly and steadily making sure that the engine does not get caught in the compartment.

▶ Before replacing the engine, check the alignment of the clutch centre plate and spigot (pilot) bush or bearing in the crankshaft. Check the engine to clutch housing locating dowels are in place. Select first or second gear in the gearbox. To replace, carefully lower the engine into the compartment and push onto the gearbox. Keep the engine and gearbox square and centralized. Rotate the engine crankshaft to align the clutch and gearbox input shaft splines.

▶ Push the engine fully home onto the locating dowels and fit two bolts to clutch housing and tighten. Remove the jack from under the gearbox and lower engine into the mountings. Fit and tighten engine mounting nuts/bolts and remove crane and sling. Replace all components in reverse order. Make a systematic check of all components for correct fitting and security. Road test and recheck under bonnet.

Job card

Technician/learner name & date	Make and model	VIN no.		Reg. no.	Job/task no.

Customer's instructions/vehicle fault	Mileage	

Work carried out and recommendations (include PPE & special precautions taken)

Parts and labour	Price
Total	

Data and specifications used (include the actual figures)

Assessor report

	Assessment outcome	*Passed (tick ✓)*
1	The learner worked safely and minimised risks to themselves and others	
2	The learner correctly selected and used appropriate technical information	
3	The learner correctly selected and used appropriate tools and equipment	
4	The learner correctly carried out the task required using suitable methods and testing procedures	
5	The learner correctly recorded information and made suitable recommendations	

	Tick	Written feedback (with reference to assessment criteria) must be given when a learner is referred
Pass: I confirm that the learner's work was to an acceptable standard and met the assessment criteria of the unit		
Refer: The work carried out did not achieve the standards specified by the assessment criteria		

Assessor name (print)	Assessor PIN/ref.	Date

Section below only to be completed by the learner once the assessor decision has been made and feedback given

I confirm that the work carried out was my own, and that I received feedback from the Assessor	Learner name (print)	Learner signature	Date

Worksheet 6: Remove and reinstall transverse engine and transmission (FWD) with crane

Procedure

▶ Obtain crane, sling and lifting eyes with a SWL in excess of engine weight. Disconnect battery earth/ground cable. Fit lifting eyes to engine as per manufacturer's instructions.

▶ Label and disconnect all electrical cables to engine including engine ground cable. Label and disconnect all vacuum pipes to engine. Remove air intake ducting and air cleaner, blank off open pipes. Depressurize, label and disconnect fuel lines to engine, blank off pipes. Label and disconnect throttle/choke and other control cables. Disconnect exhaust down pipe at manifold flange.

▶ Disconnect the gear change remote linkage at the gearbox. Remove the speedo drive cable. Disconnect the clutch cable or hydraulic cylinder complete with pipes. Remove the drive shafts from the final drive (catch oil). Fit dummy shafts to retain sun wheels. Slacken and remove, if safe to do so, the engine mounting nuts from the underside of vehicle.

▶ When all underside work is completed, drain the coolant and remove radiator and all hoses. When air conditioning is fitted, do not disconnect refrigerant pipes/hoses before removing refrigerant (specialist task). If possible, remove and support the compressor with hoses attached.

▶ Attach sling to lifting eyes and crane, tension sling. Check all connections to engine and gearbox have been disconnected. Remove engine mounting nuts/bolts if not already removed. Operate crane to lift engine clear of mountings. Lift slowly and steadily making sure that the engine and gearbox do not get caught in the compartment.

▶ To replace, carefully lower the engine and gearbox into the compartment and locate onto the engine mountings. Fit engine mounting nuts/bolts and tighten. Remove crane and sling. Refit drive shafts and check and top up gearbox oil. Replace all components in reverse order. Make a systematic check of all components for correct fitting and security. Start and run engine, bleed cooling system, if necessary, check for correct operation. Road test and recheck under bonnet.

Job card

Technician/learner name & date	Make and model	VIN no.		Reg. no.	Job/task no.

Customer's instructions/vehicle fault		Mileage	

Work carried out and recommendations (include PPE & special precautions taken)

Parts and labour	Price
Total	

Data and specifications used (include the actual figures)

Assessor report

	Assessment outcome	Passed (tick ✓)
1	The learner worked safely and minimised risks to themselves and others	
2	The learner correctly selected and used appropriate technical information	
3	The learner correctly selected and used appropriate tools and equipment	
4	The learner correctly carried out the task required using suitable methods and testing procedures	
5	The learner correctly recorded information and made suitable recommendations	

	Tick	Written feedback (with reference to assessment criteria) must be given when a learner is referred
Pass: I confirm that the learner's work was to an acceptable standard and met the assessment criteria of the unit		
Refer: The work carried out did not achieve the standards specified by the assessment criteria		

Assessor name (print)	Assessor PIN/ref.	Date

Section below only to be completed by the learner once the assessor decision has been made and feedback given			
I confirm that the work carried out was my own, and that I received feedback from the Assessor	Learner name (print)	Learner signature	Date

Worksheet 7: Remove and reinstall transverse engine and transmission onto trolley (includes front frame and suspension, OBD1 or newer)

Procedure

▶ Place vehicle on a suitable lift with the lifting points correctly positioned (see workshop manual). Obtain special or suitable trolley.

▶ Disconnect battery ground cable. Underneath the vehicle remove exhaust down pipe, gear change linkage, speedo cable, clutch control cable or hydraulics, steering column coupling and brake connections (hoses or pipes). Blank off pipes.

▶ Under the bonnet label and disconnect electrical terminals, vacuum connections, fuel feed and return pipes (plug or cap pipes), air ducting, throttle and other control cables. Drain coolant and remove radiator and hoses.

▶ When air conditioning is fitted, do not disconnect refrigerant pipes/hoses before removing refrigerant (specialist task). If possible, remove and support the compressor with hoses attached.

▶ Check all connections to the engine, transmission and sub frame have been disconnected.

▶ Lower vehicle so that the sub frame rests on the trolley. Undo and remove sub frame to chassis mounting bolts.

Note: check that bolts are not supporting a load before removal.

▶ Carefully lift the vehicle away from the sub frame, watching that the engine and transmission does not catch in the engine compartment.

▶ Refit and reassemble in the reverse order. Tighten all nuts/bolts to specified torque settings. Make a systematic check of all components for correct fitting and security.

▶ Bleed and check brakes. Check front wheel alignment. Run the engine, bleed cooling system, if necessary, and check for correct operation.

▶ Road test and recheck all work under the vehicle and under the bonnet.

Job card

Technician/learner name & date	Make and model	VIN no.	Reg. no.	Job/task no.

Customer's instructions/vehicle fault		Mileage		

Work carried out and recommendations (include PPE & special precautions taken)

Parts and labour	Price
Total	

Data and specifications used (include the actual figures)

Assessor report

	Assessment outcome	Passed (tick ✓)
1	The learner worked safely and minimised risks to themselves and others	
2	The learner correctly selected and used appropriate technical information	
3	The learner correctly selected and used appropriate tools and equipment	
4	The learner correctly carried out the task required using suitable methods and testing procedures	
5	The learner correctly recorded information and made suitable recommendations	

	Tick	Written feedback (with reference to assessment criteria) must be given when a learner is referred
Pass: I confirm that the learner's work was to an acceptable standard and met the assessment criteria of the unit		
Refer: The work carried out did not achieve the standards specified by the assessment criteria		

Assessor name (print)	Assessor PIN/ref.	Date

Section below only to be completed by the learner once the assessor decision has been made and feedback given			
I confirm that the work carried out was my own, and that I received feedback from the Assessor	Learner name (print)	Learner signature	Date

Worksheet 8: Inspect and replace sump, covers, gaskets and seals

Procedure

▶ Disconnect battery ground cable.

▶ Paper gaskets – remove cover/housing and clean old gasket material and sealant from both gasket faces. Check faces for flatness, rectify as necessary. Fit new gasket with sealant, if specified, refit bolts and tighten to specified torque.

▶ Rubber gaskets – remove cover/housing and clean old gasket material from both faces. Check gasket faces for flatness, rectify as necessary. Fit new gasket without sealant to clean and dry faces, refit bolts and tighten until gasket is pinched. Do not over tighten.

▶ Cork gaskets – remove cover/sump and clean old gasket material and sealant from both gasket faces. Check faces for flatness, rectify as necessary. Fit new gasket with sealant, if specified, refit bolts and tighten to specified torque and in sequence.

▶ Formed in place gaskets (tube applied) – remove cover/sump and clean old gasket sealant from both faces. Check faces for flatness, rectify as necessary. Apply sealant bead gaskets as per manufacturer's instructions. Fit and torque in sequence within 3 minutes.

▶ Oil seals – these are used with gaskets for sealing around bearing caps and oil moulded rubber pans. Fit with recommended sealants or dry and then fit other type of gaskets and sump. Fit and torque bolts in sequence. Check seals do not squeeze out.

▶ Oil seals (lip) – remove pulley from housing and inspect seal land. Use a special tool to extract the seal. Press in new seal to shoulder or specified depth. Lubricate (in block) seal with clean engine oil. Clean and lubricate seal land, refit pulley and torque bolt.

Job card

Technician/learner name & date	Make and model	VIN no.	Reg. no.	Job/task no.

Customer's instructions/vehicle fault		Mileage	

Work carried out and recommendations (include PPE & special precautions taken)

Parts and labour	Price
Total	

Data and specifications used (include the actual figures)

Assessor report

	Assessment outcome	Passed (tick ✓)
1	The learner worked safely and minimised risks to themselves and others	
2	The learner correctly selected and used appropriate technical information	
3	The learner correctly selected and used appropriate tools and equipment	
4	The learner correctly carried out the task required using suitable methods and testing procedures	
5	The learner correctly recorded information and made suitable recommendations	

	Tick	Written feedback (with reference to assessment criteria) must be given when a learner is referred
Pass: I confirm that the learner's work was to an acceptable standard and met the assessment criteria of the unit		
Refer: The work carried out did not achieve the standards specified by the assessment criteria		

Assessor name (print)	Assessor PIN/ref.	Date

Section below only to be completed by the learner once the assessor decision has been made and feedback given			
I confirm that the work carried out was my own, and that I received feedback from the Assessor	Learner name (print)	Learner signature	Date

Worksheet 9: Remove and replace crankshaft, vibration damper, flywheel and clutch pilot bush/bearing

Procedure

▶ Remove gearbox and clutch, strip underside of engine, sump and front cover to access crankshaft. Check and record the end float on the crankshaft. Undo front pulley/vibration damper bolt (lock flywheel to aid undoing of bolt). Pull off pulley/damper and inspect seal land and pulley vee faces. Remove the front cover and timing belt or chain. Check and mark direction of rotation and timing. Remove bolts securing the flywheel/flex plate/drive plate and carefully remove flywheel taking weight in both hands. Remove rear main oil seal housing. Inspect flywheel.

▶ Check the big end and main bearing caps for position (No. 1, 2, etc.) and slide to the front of engine. If not marked, mark with dots with a centre punch or similar. Remove big end bearing caps and lay out in order. Remove main bearing caps and lay out in order. Lift out crankshaft. If working below vehicle, remove centre bearing and crankshaft together. Inspect crankshaft for surface cracks and journal damage. Remove all bearing shells if they are to be replaced. Check oil passage condition. Measure journal wear. Fit new bearings into connecting rods and main bearings caps, locate tangs and oil way holes. Check new bearings for 'nip' by assembling finger tight and checking cap gap with feeler gauge.

▶ Pull out clutch pilot bearing from crankshaft. If a bush, fill with grease and punch in with a suitable mandrel to push out bush. Otherwise use special puller. Check new bearing fits gearbox input shaft and then fit into crankshaft. There is no bush/bearing on some transaxles and auto gearboxes.

▶ Fit new oil seals to front and rear housings or around crankshaft if appropriate. Lubricate all bearings and journals with clean engine oil. Fit crankshaft into main bearings in the block. Check end float. Fit main bearing caps and tighten the bolts to the specified torque and check engine rotation. Fit big end bearing caps and tighten to torque and check engine rotation (fit timing chain). Fit front and rear main covers, centralize and tighten bolts in sequence and to torque. Fit flywheel/ drive plate to dowels and fit bolts, apply sealant or thread locking compound, if specified. Tighten bolts in sequence and to torque. Check run out. Fit timing belt and pulleys and covers. Fit pulley/ vibration damper. Before fitting new oil filter, *prime* the pump with clean engine oil through the oil feed hole to the oil filter. Fit filter. Make a systematic check of all components for correct fitting and security. Fit gearbox, sump/sump and other components. Fill engine with oil, start and run engine and check for correct operation.

Job card

Technician/learner name & date	Make and model	VIN no.		Reg. no.	Job/task no.

Customer's instructions/vehicle fault		Mileage	

Work carried out and recommendations (include PPE & special precautions taken)

Parts and labour	Price
Total	

Data and specifications used (include the actual figures)

Assessor report

	Assessment outcome	Passed (tick ✓)
1	The learner worked safely and minimised risks to themselves and others	
2	The learner correctly selected and used appropriate technical information	
3	The learner correctly selected and used appropriate tools and equipment	
4	The learner correctly carried out the task required using suitable methods and testing procedures	
5	The learner correctly recorded information and made suitable recommendations	

	Tick	Written feedback (with reference to assessment criteria) must be given when a learner is referred
Pass: I confirm that the learner's work was to an acceptable standard and met the assessment criteria of the unit		
Refer: The work carried out did not achieve the standards specified by the assessment criteria		

Assessor name (print)	Assessor PIN/ref.	Date

Section below only to be completed by the learner once the assessor decision has been made and feedback given			
I confirm that the work carried out was my own, and that I received feedback from the Assessor	Learner name (print)	Learner signature	Date

Worksheet 10: Inspect crankshaft condition, journal and bearing dimensions and wear

Procedure

▶ Strip to access the crankshaft. Check end float and general condition before removing. Check all bearing caps are marked for position and location to the front of the engine.

▶ Remove crankshaft and bearings. Visually inspect bearings for condition, wear and abnormal markings.

▶ Visually inspect the crankshaft journals and thrust faces for scores, pitting and abnormal wear or uneven markings.

▶ Lay the crankshaft in oiled paper supports in vee blocks on a surface table. Set up a dial test indicator (DTI) on a base. Measure the crankshaft for bow (support ends – rotate and measure centre) and twist (compare matched pairs of journals, 1&4, 2&3, etc.).

▶ Measure and record all journal sizes for nominal size, taper and ovality. Compare with manufacturer's data for wear tolerances.

▶ Check oil ways in the crankshaft are clear and clean. Flush through with low pressure jet on cleaning tank.

▶ If crankshaft is reground, wash thoroughly before refitting. Fit correct bearings (undersize if crankshaft has been reground). Check on journals before fitting.

▶ Lubricate all bearings with clean engine oil. Check 'nip' of bearings before tightening. Torque bearing cap bolts to specified torque and check engine rotates. Check 'feel' of bearings and end float of crankshaft before reassembling engine.

▶ When the engine is running check for abnormal noises and correct operation.

Job card

Technician/learner name & date	Make and model	VIN no.	Reg. no.	Job/task no.

Customer's instructions/vehicle fault	Mileage	

Work carried out and recommendations (include PPE & special precautions taken)

Parts and labour	Price
Total	

Data and specifications used (include the actual figures)

Assessor report

	Assessment outcome	Passed (tick ✓)
1	The learner worked safely and minimised risks to themselves and others	
2	The learner correctly selected and used appropriate technical information	
3	The learner correctly selected and used appropriate tools and equipment	
4	The learner correctly carried out the task required using suitable methods and testing procedures	
5	The learner correctly recorded information and made suitable recommendations	

	Tick	Written feedback (with reference to assessment criteria) must be given when a learner is referred
Pass: I confirm that the learner's work was to an acceptable standard and met the assessment criteria of the unit		
Refer: The work carried out did not achieve the standards specified by the assessment criteria		

Assessor name (print)	Assessor PIN/ref.	Date

Section below only to be completed by the learner once the assessor decision has been made and feedback given

I confirm that the work carried out was my own, and that I received feedback from the Assessor	Learner name (print)	Learner signature	Date

Worksheet 11: Inspect pistons, piston rings and connecting rods and bearings

Procedure

▶ Remove cylinder head and sump to access pistons and connecting rods. Before removing pistons check the condition of the piston crowns and the 'fit' of the pistons in the cylinder bores. Rock the pistons from side to side to 'feel' the free play.

▶ Visually inspect cylinder bores for scores, indicating broken piston rings, and for uneven wear to one side, indicating misalignment or a bent connecting rod. Before removing the connecting rod big end bearing caps check for free play by 'feel' by pulling and pushing the connecting rod on the crankshaft journal. Check end float by side to side movement and feeler gauge. Compare with manufacturer's data.

▶ Remove connecting rods and pistons. Visually inspect the big end bearings for uneven wear indicating a possible bent connecting rod. Visually inspect the cylinder bore for matched wear. Visually inspect the crankshaft journals and bearings. If the bearings are worn, measure the crankshaft journals and compare with the manufacturer's specifications and tolerances for fitting new bearings without regrinding the crankshaft.

▶ Clean the piston crown and inspect for burning or overheating damage, pitting, particle damage such as broken piston rings and other deterioration.

▶ Visually check and measure piston ring grooves and ring land for condition. Look for opening or taper of the ring grooves caused by the rings rocking. Look for ring breakage and particles cutting through the ring land.

▶ Remove the piston from the connecting rod. Check the 'fit' of the piston pin to the connecting rod little end bearing/bush and in the piston.

▶ If the piston is serviceable and new rings are to be fitted, measure the cylinder bores and compare with the manufacturer's specifications and tolerance limits for new rings.

▶ Gap new rings in the cylinder. Clean the piston ring grooves using a special tool. Fit rings to pistons and reassemble engine. Run engine and check for abnormal noises and correct operation.

Job card

Technician/learner name & date	Make and model	VIN no.		Reg. no.	Job/task no.

Customer's instructions/vehicle fault		Mileage	

Work carried out and recommendations (include PPE & special precautions taken)

Parts and labour		Price
Total		

Data and specifications used (include the actual figures)

Assessor report

	Assessment outcome	Passed (tick ✓)
1	The learner worked safely and minimised risks to themselves and others	
2	The learner correctly selected and used appropriate technical information	
3	The learner correctly selected and used appropriate tools and equipment	
4	The learner correctly carried out the task required using suitable methods and testing procedures	
5	The learner correctly recorded information and made suitable recommendations	

	Tick	Written feedback (with reference to assessment criteria) must be given when a learner is referred
Pass: I confirm that the learner's work was to an acceptable standard and met the assessment criteria of the unit		
Refer: The work carried out did not achieve the standards specified by the assessment criteria		

Assessor name (print)	Assessor PIN/ref.	Date

Section below only to be completed by the learner once the assessor decision has been made and feedback given			
I confirm that the work carried out was my own, and that I received feedback from the Assessor	Learner name (print)	Learner signature	Date

Worksheet 12: Remove and replace cylinder heads OHV

Procedure

▶ Check permissible conditions for head removal – allow engine to cool if required. Disconnect battery ground cable.

▶ Label and remove all electrical terminals to components on cylinder head and manifolds. Label and remove all vacuum pipes/hoses to cylinder head and manifolds. Remove distributor cap and spark plug leads.

▶ Obtain a clean drain tray and drain engine coolant. Remove radiator, top, bypass and heater hoses.

▶ Remove air intake ducts, air cleaner, and PCV hoses as necessary, etc. Label and remove throttle and choke cables. Remove carburettor/monopoint injector body/inlet manifold and injectors – blank off all fuel pipes.

▶ Remove exhaust at down pipe flange or manifold to head, pull clear of cylinder head. Remove rocker cover and rocker shaft. Shake the push rods to free them from the followers. Remove push rods and store in order (drilled block or similar) so that they can be replaced in exactly the same positions.

▶ Working in the reverse order to the tightening sequence, undo the cylinder head bolts by a small amount (¼ of a turn). Fully undo and remove head bolts. Lay out in order if the bolts are reusable. Note any bolts that do not run cleanly in their threads. Clean threads. Lift off cylinder head (twist first if wet liners are fitted). Clean faces of the cylinder head and block and inspect. Plug water jacket to prevent particles of dirt entering.

▶ Fit new cylinder head gasket – follow manufacturer's instructions – use guide studs if necessary. Lower head into place on locating dowels. Fit head bolts and do up finger tight. Follow manufacturer's instructions for head bolt tightening sequence and torque settings. Refit push rods, rocker shaft and rockers; adjust valve clearances (lash). Refit rocker cover. Refit exhaust, inlet, fuel electrical and coolant components. Top up coolant. Run engine, bleed heater if necessary, recheck coolant level and coolant circulation. Listen to engine for abnormal noises and oil/coolant leaks and operation of warning lights/gauges.

▶ Road test and check operation of engine, coolant temperature, heater operation, etc. Recheck after road test inside engine compartment for oil/coolant leaks, security of all disturbed components and fittings and cleanliness – rectify and recheck if necessary.

Job card

Technician/learner name & date	Make and model	VIN no.	Reg. no.	Job/task no.

Customer's instructions/vehicle fault	Mileage	

Work carried out and recommendations (include PPE & special precautions taken)

Parts and labour	**Price**
Total	

Data and specifications used (include the actual figures)

Assessor report

Assessment outcome		*Passed (tick ✓)*
1	The learner worked safely and minimised risks to themselves and others	
2	The learner correctly selected and used appropriate technical information	
3	The learner correctly selected and used appropriate tools and equipment	
4	The learner correctly carried out the task required using suitable methods and testing procedures	
5	The learner correctly recorded information and made suitable recommendations	

	Tick	Written feedback (with reference to assessment criteria) must be given when a learner is referred
Pass: I confirm that the learner's work was to an acceptable standard and met the assessment criteria of the unit		
Refer: The work carried out did not achieve the standards specified by the assessment criteria		

Assessor name (print)	Assessor PIN/ref.	Date

Section below only to be completed by the learner once the assessor decision has been made and feedback given			
I confirm that the work carried out was my own, and that I received feedback from the Assessor	Learner name (print)	Learner signature	Date

Worksheet 13: Remove and replace cylinder heads OHC

Procedure

▶ Check permissible conditions for head removal – allow engine to cool if required. Disconnect battery ground cable. Label and remove all electrical terminals to components on cylinder head and manifolds. Label and remove all vacuum pipes/hoses to cylinder head and manifolds. Remove distributor cap and spark plug leads. Obtain a clean drain tray and drain engine coolant. Remove radiator, top, bypass and heater hoses. Remove air intake ducts, air cleaner, and PCV hoses as necessary, etc. Label and remove throttle and choke cables. Remove carburettor/monopoint injector body/inlet manifold and injectors – blank off all fuel pipes.

▶ Remove exhaust at down pipe flange or manifold to head, pull clear of cylinder head. Remove cam housing cover and gasket. Remove front/camshaft drive belt cover – mark direction of rotation (DOR) on belt and turn engine to align timing marks. Slacken tensioner and remove belt. *Do not turn* the crankshaft or camshaft before the head is removed. Return to this position before the head is replaced. Working in the reverse order to the tightening sequence, undo the cylinder head bolts by a small amount (¼ of a turn). Fully undo and remove head bolts. Lay out in order if the bolts are reusable. Note any bolts that do not run cleanly in their threads. Clean threads. Lift off cylinder head (twist first if wet liners are fitted). Clean faces of the cylinder head and block and inspect. Plug water jacket to prevent particles of dirt entering.

▶ Fit new cylinder head gasket – follow manufacturer's instructions – use guide studs if necessary. Check and position crankshaft and camshaft to align timing marks. Lower head into place on locating dowels. Fit head bolts and do up finger tight. Follow manufacturer's instructions for head bolt tightening sequence and torque settings. Refit camshaft drive belt, check timing marks align; tension belt and rotate for two or three revolutions, recheck timing and belt tension. Fit cam housing cover with new gasket.

▶ Refit exhaust, inlet, fuel electrical and coolant components. Top up coolant. Run engine, bleed heater if necessary, recheck coolant level and coolant circulation. Listen to engine for abnormal noises and oil/coolant leaks and operation of warning lights/gauges.

▶ Road test and check operation of engine, coolant temperature, heater operation, etc. Recheck after road test inside engine compartment for oil/coolant leaks, security of all disturbed components and fittings and cleanliness – rectify and recheck if necessary.

Job card

Technician/learner name & date	Make and model	VIN no.		Reg. no.	Job/task no.

Customer's instructions/vehicle fault		Mileage			

Work carried out and recommendations (include PPE & special precautions taken)

	Price
Parts and labour	
Total	

Data and specifications used (include the actual figures)

Assessor report

Assessment outcome		Passed (tick ✓)
1	The learner worked safely and minimised risks to themselves and others	
2	The learner correctly selected and used appropriate technical information	
3	The learner correctly selected and used appropriate tools and equipment	
4	The learner correctly carried out the task required using suitable methods and testing procedures	
5	The learner correctly recorded information and made suitable recommendations	

	Tick	Written feedback (with reference to assessment criteria) must be given when a learner is referred
Pass: I confirm that the learner's work was to an acceptable standard and met the assessment criteria of the unit		
Refer: The work carried out did not achieve the standards specified by the assessment criteria		

Assessor name (print)	Assessor PIN/ref.	Date

Section below only to be completed by the learner once the assessor decision has been made and feedback given			
I confirm that the work carried out was my own, and that I received feedback from the Assessor	Learner name (print)	Learner signature	Date

Worksheet 14: Remove and replace cam followers including hydraulic tappets

Procedure

▶ Run engine and listen for abnormal noises from valve mechanism. Strip engine to access cam followers – in side of engine (OHV) or in cylinder head (OHC).

▶ Alternatives – remove push rods and side covers – lift out followers. OHV – remove cylinder head and lift out followers – remove inlet manifold and covers (vee engines) and lift out followers.

▶ Alternatives – cam in head and rocker actuation – remove rockers which are cam followers OHC – direct acting cams require removal of the camshaft to access cam followers.

▶ Check cam followers before removal, for free play across the bore and that the follower is free to rotate.

▶ Lift out followers and lay out in order (they must be replaced in exactly the same position). Check thrust faces and sides of cam followers and the cam lobes. Inspect for signs that indicate that the follower has or has not been rotating.

▶ For hydraulic tappets/followers check oil feed holes in the block/head/tappet are clean. Follow the manufacturer's instructions for stripping and inspecting hydraulic tappets.

▶ Lubricate all parts before and during assembly with clean engine oil. Replace followers/tappets in correct position. Reassemble other components in reverse order. Check and adjust valve clearances.

▶ Refit any other parts and battery ground cable.

▶ Start and run engine – check for correct operation.

▶ Road test and recheck.

Job card

Technician/learner name & date	Make and model	VIN no.		Reg. no.	Job/task no.

Customer's instructions/vehicle fault	Mileage	

Work carried out and recommendations (include PPE & special precautions taken)	

Parts and labour	Price
Total	

Data and specifications used (include the actual figures)

Assessor report

	Assessment outcome	Passed (tick ✓)
1	The learner worked safely and minimised risks to themselves and others	
2	The learner correctly selected and used appropriate technical information	
3	The learner correctly selected and used appropriate tools and equipment	
4	The learner correctly carried out the task required using suitable methods and testing procedures	
5	The learner correctly recorded information and made suitable recommendations	

	Tick	Written feedback (with reference to assessment criteria) must be given when a learner is referred
Pass: I confirm that the learner's work was to an acceptable standard and met the assessment criteria of the unit		
Refer: The work carried out did not achieve the standards specified by the assessment criteria		

Assessor name (print)	Assessor PIN/ref.	Date

Section below only to be completed by the learner once the assessor decision has been made and feedback given			
I confirm that the work carried out was my own, and that I received feedback from the Assessor	Learner name (print)	Learner signature	Date

Worksheet 15: Perform cylinder compression tests; determine necessary action

Procedure

▶ Check engine manufacturer's data for suitability of this test and checking conditions (cold or hot). Disconnect LT (low tension) or primary circuit feed at coil to prevent HT current. (LT feed from coil to distributor).

▶ Remove all spark plugs.

▶ Connect compression tester into No. 1 spark plug hole. Fully open throttle.

▶ Crank engine for three or four revolutions, record initial (first revolution) and final compression reading. (Seek assistance if necessary to hold the throttle and to crank the engine.)

▶ Repeat for all other cylinders and record readings.

▶ If some readings are low, put a small drop of clean engine oil into the cylinder and repeat test. Record results.

▶ Repeat for all cylinders.

▶ Compare dry and wet readings with other cylinders and with the engine manufacturer's specifications.

▶ Refit spark plugs, HT leads and coil LT lead – check engine starts and runs.

Job card

Technician/learner name & date	Make and model	VIN no.	Reg. no.	Job/task no.

Customer's instructions/vehicle fault		Mileage		

Work carried out and recommendations (include PPE & special precautions taken)

Parts and labour	Price
Total	

Data and specifications used (include the actual figures)

Assessor report

	Assessment outcome	Passed (tick ✓)
1	The learner worked safely and minimised risks to themselves and others	
2	The learner correctly selected and used appropriate technical information	
3	The learner correctly selected and used appropriate tools and equipment	
4	The learner correctly carried out the task required using suitable methods and testing procedures	
5	The learner correctly recorded information and made suitable recommendations	

	Tick	Written feedback (with reference to assessment criteria) must be given when a learner is referred
Pass: I confirm that the learner's work was to an acceptable standard and met the assessment criteria of the unit		
Refer: The work carried out did not achieve the standards specified by the assessment criteria		

Assessor name (print)	Assessor PIN/ref.	Date

Section below only to be completed by the learner once the assessor decision has been made and feedback given			
I confirm that the work carried out was my own, and that I received feedback from the Assessor	Learner name (print)	Learner signature	Date

Worksheet 16: Remove and replace oil pumps and drive mechanisms – time oil pump and distributor shaft

Procedure

▶ *Oil pump drives distributor.* If the oil pump and the distributor share the same drive from the camshaft, position the engine so that No. 1 cylinder is at TDC on the compression stroke. Remove the distributor cap and mark or record the position of the distributor and rotor arm. Remove the distributor and record the position of the drive dogs (large 'D' position). Replaced oil pump must return to this position when reassembled.

▶ On the underside of the engine disconnect and remove the oil pick up pipe and strainer. Undo and remove the oil pump securing bolts and remove the pump. Clean old gasket from mating faces. Strip and inspect pump. Refit pump with new gasket or 'O' ring and sealant *only* if specified. Turn drive gear to the timing position and ease home on the camshaft gear. Check the 'D' drives dog position in the distributor-housing bore. Correct if necessary by pulling back and turning the gear appropriately. When the distributor drive is correctly positioned fit and tightens the securing bolts to the specified torque.

▶ Fit the pickup pipe and strainer using new seals and tighten securing bolts. Refit the distributor and set ignition timing. Finally adjust timing when the engine is running.

▶ *Oil pump not timed.* If the oil pump is driven from the crankshaft or camshaft or an auxiliary shaft independently of the ignition system, the pump drive mechanism need not be timed. Remove pickup pipe and strainer if necessary. Undo oil pump securing bolts and remove. Clean old gaskets or 'O' rings from pump and engine. Strip and inspect oil pump. Prime new pump with clean engine oil. Fit new gaskets or 'O' rings and sealant *only* if specified. Turn drive to locate with drive gear, dog or 'D' drive and fit pump. Fit and tighten securing bolts to specified torque. Reassemble engine and top up with engine oil. Before fitting new oil filter, *prime* the pump with clean engine oil through the oil feed hole to the oil filter. Fit filter. Start and run engine – if oil warning light does not go out in normal time stop the engine and investigate. Look under the engine for oil leaks as soon as the oil warning light goes out.

Job card

Technician/learner name & date	Make and model	VIN no.		Reg. no.	Job/task no.

Customer's instructions/vehicle fault		Mileage	

Work carried out and recommendations (include PPE & special precautions taken)

Parts and labour	Price
Total	

Data and specifications used (include the actual figures)

Assessor report

Assessment outcome		*Passed (tick ✓)*
1	The learner worked safely and minimised risks to themselves and others	
2	The learner correctly selected and used appropriate technical information	
3	The learner correctly selected and used appropriate tools and equipment	
4	The learner correctly carried out the task required using suitable methods and testing procedures	
5	The learner correctly recorded information and made suitable recommendations	

	Tick	Written feedback (with reference to assessment criteria) must be given when a learner is referred
Pass: I confirm that the learner's work was to an acceptable standard and met the assessment criteria of the unit		
Refer: The work carried out did not achieve the standards specified by the assessment criteria		

Assessor name (print)	Assessor PIN/ref.	Date

Section below only to be completed by the learner once the assessor decision has been made and feedback given			
I confirm that the work carried out was my own, and that I received feedback from the Assessor	Learner name (print)	Learner signature	Date

Worksheet 17: Remove and replace oil filters, oil coolers and turbo chargers

Procedure

▶ Allow engine to stand and cool down before starting work. Components can be very hot. Disconnect battery ground lead.

▶ Place a drain tray below work area to catch oil. Use a small tray/container to catch oil immediately below work area.

▶ Many oil feed pipes and hoses are held with a union nut. Use two spanners, one to hold the pipe or hose or connector and the other to undo the union nut. Undo the highest point first to allow the pipe/hose to drain before removal.

▶ Undo the lowest point and catch lost oil and withdraw the pipe/hose. *Avoid* bending or distorting pipes.

▶ Drain coolant from cooling system to oil circuit heat exchangers and disconnect hoses. Undo securing bolts on all types of oil coolers/heat exchangers and remove.

▶ Refit oil coolers/heat exchangers, refit and tighten securing bolts.

▶ Refit pipes/hoses by assembling both ends before tightening. Check the alignment and routing of the pipe/hose before tightening the union nuts. Use two spanners, one to hold the pipe, hose or connector and the other to do up the union nut.

▶ Check and top up oil (and coolant) level.

▶ Run engine and check for oil (coolant) leaks.

Job card

Technician/learner name & date	Make and model	VIN no.	Reg. no.	Job/task no.

Customer's instructions/vehicle fault	Mileage	

Work carried out and recommendations (include PPE & special precautions taken)	

Parts and labour	Price
Total	

Data and specifications used (include the actual figures)

Assessor report

Assessment outcome		Passed (tick ✓)
1	The learner worked safely and minimised risks to themselves and others	
2	The learner correctly selected and used appropriate technical information	
3	The learner correctly selected and used appropriate tools and equipment	
4	The learner correctly carried out the task required using suitable methods and testing procedures	
5	The learner correctly recorded information and made suitable recommendations	

	Tick	Written feedback (with reference to assessment criteria) must be given when a learner is referred
Pass: I confirm that the learner's work was to an acceptable standard and met the assessment criteria of the unit		
Refer: The work carried out did not achieve the standards specified by the assessment criteria		

Assessor name (print)	Assessor PIN/ref.	Date

Section below only to be completed by the learner once the assessor decision has been made and feedback given			
I confirm that the work carried out was my own, and that I received feedback from the Assessor	Learner name (print)	Learner signature	Date

Worksheet 18: Remove and replace oil pressure switch and oil pressure gauge sender units

Procedure

▶ Allow engine to stand and cool down before starting work. Components can be very hot. Disconnect battery ground lead.

▶ Place a drain tray below work area to catch oil. Use a small tray/container to catch oil immediately below work area.

▶ Disconnect the electrical feed to the oil pressure switch or electronic sensor. Undo the switch using a spanner on the hexagon at the end of the threaded insert to the oil gallery.

▶ Carry out a pressure test if low oil pressure is suspected. This tests would be sensible but not required if the oil switch were being replaced because of leakage.

▶ Refit by screwing in the new switch or sensor until it is tight in the tapered thread or tight on a new sealing washer.

▶ Reconnect the battery and run the engine to check the operation of the warning light or pressure gauge. Check the switch for oil leaks.

▶ Check and top up oil level.

▶ The oil level sensor will be located in the engine block or sump. Look up and follow the manufacturer's instructions for this task.

Job card

Technician/learner name & date	Make and model	VIN no.	Reg. no.	Job/task no.

Customer's instructions/vehicle fault	Mileage	

Work carried out and recommendations (include PPE & special precautions taken)

Parts and labour	Price
Total	

Data and specifications used (include the actual figures)

Assessor report

	Assessment outcome	Passed (tick ✓)
1	The learner worked safely and minimised risks to themselves and others	
2	The learner correctly selected and used appropriate technical information	
3	The learner correctly selected and used appropriate tools and equipment	
4	The learner correctly carried out the task required using suitable methods and testing procedures	
5	The learner correctly recorded information and made suitable recommendations	

	Tick	Written feedback (with reference to assessment criteria) must be given when a learner is referred
Pass: I confirm that the learner's work was to an acceptable standard and met the assessment criteria of the unit		
Refer: The work carried out did not achieve the standards specified by the assessment criteria		

Assessor name (print)	Assessor PIN/ref.	Date

Section below only to be completed by the learner once the assessor decision has been made and feedback given			
I confirm that the work carried out was my own, and that I received feedback from the Assessor	Learner name (print)	Learner signature	Date

Worksheet 19: Inspect cooling system, pressure test, check coolant condition and antifreeze

Procedure

▶ Obtain a suitable cooling system pressure tester and an ethylene glycol coolant hydrometer. Remove the radiator or expansion tank cap. Check the coolant for level, colour and contamination such as oil or rust and dirt.

▶ Check the ethylene glycol content with the hydrometer – adjust for temperature – and compare with manufacturer's specifications (usually between 25% and 50% depending on the area of vehicle operation).

▶ Top up the coolant if necessary and attach the pressure tester using a suitable adapter in the filler neck.

▶ Apply a pressure equal to the system operating pressure, which is shown on the radiator filler cap – but should be checked from the manufacturer's data. *Do not* exceed the stated pressure. Observe the reading on the tester pressure gauge – it should remain stable. If the gauge reading drops, look for a good connection to the system and then for leaks in the system.

▶ Look at all hoses, pipes, joints, gaskets, the water pump, the heater and water valve for external leaks. Look inside the vehicle under the heater for leaks from the heater matrix. If no external leaks are visible, check the coolant for oil contamination (possible internal leaks.) Apply a pressure of about half the operating pressure and run the engine. Look for a rapid pressure rise, which would indicate a cylinder head gasket leak or crack between the cylinder or cylinder head and the water jacket.

▶ If the pressure remains stable but internal leakage is suspected, use a combustion gas detector kit with the pressure tester. If the test fluid changes colour during the test, combustion gases have been detected in the coolant confirming an internal leak.

▶ Fit the radiator/expansion tank pressure cap to the tester using an appropriate adapter and check that the operating pressure is held at the specified value. Remove the cap and check the vacuum valve seal operation and condition. Compare all findings with manufacturer's data. Report faults found.

Job card

Technician/learner name & date	Make and model	VIN no.		Reg. no.	Job/task no.
Customer's instructions/vehicle fault		Mileage			
Work carried out and recommendations (include PPE & special precautions taken)					
Parts and labour					Price
Total					
Data and specifications used (include the actual figures)					

Assessor report

	Assessment outcome	Passed (tick ✓)
1	The learner worked safely and minimised risks to themselves and others	
2	The learner correctly selected and used appropriate technical information	
3	The learner correctly selected and used appropriate tools and equipment	
4	The learner correctly carried out the task required using suitable methods and testing procedures	
5	The learner correctly recorded information and made suitable recommendations	

	Tick	Written feedback (with reference to assessment criteria) must be given when a learner is referred
Pass: I confirm that the learner's work was to an acceptable standard and met the assessment criteria of the unit		
Refer: The work carried out did not achieve the standards specified by the assessment criteria		

Assessor name (print)	Assessor PIN/ref.	Date

Section below only to be completed by the learner once the assessor decision has been made and feedback given			
I confirm that the work carried out was my own, and that I received feedback from the Assessor	Learner name (print)	Learner signature	Date

Worksheet 20: Remove and replace drive belts and pulleys

Procedure

▶ Disconnect battery earth or ground lead. Identify the drive belt for the water pump – V, serpentine or toothed belt for the camshaft drive.

▶ For water pumps driven by the camshaft drive toothed belt, follow the manufacturer's instructions to access the belt. Match the timing marks or fit the locating dowels or pegs and mark the direction of rotation on the belt before removing.

▶ For timed and non-timed belts, slacken the tensioner and remove the belt. The tensioner may be a tensioning pulley or may be linked to a component such as the alternator. Remove any other component that prevents removal of the belt. Lift off and remove the belt.

▶ Fit the new belt and adjust the tension to the manufacturer's specification. For vee belts the correct tension is usually about 13 mm or ½ inch of free play on the longest side. For toothed belts use a tension gauge or adjust so that the belt can be twisted through 90° on the long side.

▶ Tighten the tensioner securing bolts. Refit other components in reverse order. Run the engine and check for abnormal noises. A whine would indicate that the belt tension is too tight and a slapping noise of the belt against the cover indicates that the belt is too loose.

Job card

Technician/learner name & date	Make and model	VIN no.	Reg. no.	Job/task no.

Customer's instructions/vehicle fault		Mileage		

Work carried out and recommendations (include PPE & special precautions taken)

Parts and labour	Price
Total	

Data and specifications used (include the actual figures)

Assessor report

	Assessment outcome	Passed (tick ✓)
1	The learner worked safely and minimised risks to themselves and others	
2	The learner correctly selected and used appropriate technical information	
3	The learner correctly selected and used appropriate tools and equipment	
4	The learner correctly carried out the task required using suitable methods and testing procedures	
5	The learner correctly recorded information and made suitable recommendations	

	Tick	Written feedback (with reference to assessment criteria) must be given when a learner is referred
Pass: I confirm that the learner's work was to an acceptable standard and met the assessment criteria of the unit		
Refer: The work carried out did not achieve the standards specified by the assessment criteria		

Assessor name (print)	Assessor PIN/ref.	Date

Section below only to be completed by the learner once the assessor decision has been made and feedback given			
I confirm that the work carried out was my own, and that I received feedback from the Assessor	Learner name (print)	Learner signature	Date

Worksheet 21: Remove and replace radiator and electric fan motor and switches

Procedure

▶ Disconnect battery earth or ground lead. Drain radiator coolant into a clean drain tray. Undo and remove top, bottom and expansion tank hose clips and pull off hoses. Disconnect the electrical terminal block to the motor and switch. Leave the cowl, motor and switch in place. Remove the radiator and cowl as an assembly.

▶ For radiators with integral transmission oil cooling, clamp the feed and return hoses, undo the hose clips and pull off the hoses. Catch the lost oil in a clean drain tray. *Or* undo the union nuts on steel pipes and remove the pipes. Catch the lost oil in a drain tray.

▶ For radiators with an air conditioning radiator or condenser attached to the cooling radiator, undo the attaching screws and support the condenser. *Do not* undo the air conditioner refrigerant pipes unless the system has been discharged (specialist operation).

▶ Undo and remove the radiator securing bolts and remove with the brackets. Carefully lift out the radiator assembly. Detach the cowl, motor and fan from the radiator. Mark all bolts or screws for replacement in exactly the same places. Unscrew the switch from the radiator.

▶ Reassemble in the reverse order. Ensure that the correct bolts or screws are replaced in their original positions. Some bolts may be longer than others and can puncture the radiator if fitted incorrectly.

▶ Refill the cooling system. Top up with a correct water and antifreeze mixture to make good any spilt or lost coolant during draining. Run the engine and bleed air from the system if necessary.

▶ Road test and check the engine temperature gauge, the heater operation for hot, cold and intermediate settings. Check airflow temperature agrees with heater settings.

▶ After road testing, visually recheck the system for leaks.

▶ Allow the engine to cool and check the coolant level. Top up if necessary. Do not overfill.

Job card

Technician/learner name & date	Make and model	VIN no.		Reg. no.	Job/task no.

Customer's instructions/vehicle fault		Mileage	

Work carried out and recommendations (include PPE & special precautions taken)

Parts and labour	Price
Total	

Data and specifications used (include the actual figures)

Assessor report

	Assessment outcome	Passed (tick ✓)
1	The learner worked safely and minimised risks to themselves and others	
2	The learner correctly selected and used appropriate technical information	
3	The learner correctly selected and used appropriate tools and equipment	
4	The learner correctly carried out the task required using suitable methods and testing procedures	
5	The learner correctly recorded information and made suitable recommendations	

	Tick	Written feedback (with reference to assessment criteria) must be given when a learner is referred
Pass: I confirm that the learner's work was to an acceptable standard and met the assessment criteria of the unit		
Refer: The work carried out did not achieve the standards specified by the assessment criteria		

Assessor name (print)	Assessor PIN/ref.	Date

Section below only to be completed by the learner once the assessor decision has been made and feedback given			
I confirm that the work carried out was my own, and that I received feedback from the Assessor	Learner name (print)	Learner signature	Date

Worksheet 22: Remove and replace water pump and engine driven fan

Procedure

▶ Disconnect battery earth or ground lead. Drain the coolant into a clean drain tray. Identify the location of the water pump and drive belt arrangement. For water pumps driven by the camshaft drive toothed belt, follow the manufacturer's instructions to access the belt. Match the timing marks or fit the locating dowels or pegs and mark the direction of rotation on the belt before removing. For engine-driven fans attached to the water pump drive pulley, undo the securing bolts and pull off the fan. For some thermostatic viscous fans the hub will have to be drawn from the spindle after the water pump has been removed.

▶ For timed and non-timed belts, slacken the tensioner and remove the belt. The tensioner may be a tensioning pulley or may be linked to a component such as the alternator. Remove any other component that prevents removal of the belt. Lift off and remove the belt. Undo and remove the securing bolts for the water pump. Keep in order so that they are returned to their original positions – many vehicles have different length bolts around the water pump. Ease out the water pump. They can be tight when corrosion has bonded the pump into the engine block. Use a penetrating fluid and lever out or twist carefully to release the tightness. If very tight, check that all bolts have been removed!

▶ Check the gasket for signs of leakage. Where bolts run into the water jacket, check that the threads are clean and that the bolts run freely. Clean threads if necessary. Clean and check that the mating faces are in good condition. Refit the pump using a new gasket or seal and a soft setting sealant on the gasket and all bolts that run into the water passages. Tighten bolts to manufacturer's specifications.

▶ Refit the drive belt and adjust the tension to the manufacturer's specification. For V belts the correct tension is usually about 13 mm or ½ inch of free play on the longest side. For toothed belts use a tension gauge or adjust so that the belt can be twisted through 90° on the long side. Refill the cooling system. Top up with a correct water and antifreeze mixture to make good any spilt or lost coolant during draining. Run the engine and bleed air from the system if necessary.

▶ Road test and check the engine temperature gauge, the heater operation for hot, cold and intermediate settings. Check airflow temperature agrees with heater settings.

▶ After road testing, visually recheck the system for leaks.

▶ Allow the engine to cool and check the coolant level. Top up if necessary. Do not overfill.

Job card

Technician/learner name & date	Make and model	VIN no.		Reg. no.	Job/task no.

Customer's instructions/vehicle fault		Mileage		

Work carried out and recommendations (include PPE & special precautions taken)

Parts and labour		Price
Total		

Data and specifications used (include the actual figures)

Assessor report

	Assessment outcome	Passed (tick ✓)
1	The learner worked safely and minimised risks to themselves and others	
2	The learner correctly selected and used appropriate technical information	
3	The learner correctly selected and used appropriate tools and equipment	
4	The learner correctly carried out the task required using suitable methods and testing procedures	
5	The learner correctly recorded information and made suitable recommendations	

	Tick	Written feedback (with reference to assessment criteria) must be given when a learner is referred
Pass: I confirm that the learner's work was to an acceptable standard and met the assessment criteria of the unit		
Refer: The work carried out did not achieve the standards specified by the assessment criteria		

Assessor name (print)	Assessor PIN/ref.	Date

Section below only to be completed by the learner once the assessor decision has been made and feedback given			
I confirm that the work carried out was my own, and that I received feedback from the Assessor	Learner name (print)	Learner signature	Date

Worksheet 23: Inspect, clean or renew and gap spark plugs

Procedure

▶ Check age/mileage of spark plugs from customer records. Check manufacturer's data for correct plugs for vehicle.

▶ Check security of spark plug leads during disconnection – note and rectify any loose connections. Use an appropriate deep socket to remove the spark plugs. Note any tight threads and rectify before reassembly. Lay out plugs in order for diagnosis of plug or engine condition.

▶ Inspect plugs for condition – erosion of electrodes, carbon fouling, damaged insulation, thread condition and tightness of threaded terminals. Check sealing washer or taper seat condition.

▶ Look also for symptoms of engine or fuel system faults. Black sooty deposits indicate a rich fuel mixture, black oily deposits indicate piston ring or inlet valve stem wear, white/brown sooty deposits indicate a weak mixture and overheating. Investigate faults found, replace plugs.

▶ Plugs in good condition can be cleaned and re-gapped with a feeler gauge. All new or replaced plugs should be gapped to the manufacturer's specification before fitting.

▶ Adjust the earth/ground electrode to give a gap between the centre and earth/ground electrodes. *Do not* exert a force on the centre electrode, either lever out with a suitable tool to increase the gap or tap in on a solid object to reduce the gap. Check with a feeler or gap gauge.

▶ Lubricate the threads of plugs being fitted into aluminium cylinder heads (graphite or high-melting point grease as specified by the manufacturer). Fit into cylinder head and hand tighten before finally tightening with a torque wrench. Ensure correct torque for washer or taper seat types.

▶ Reconnect leads and check security – leads must be tight and make a good electrical connection.

▶ Run engine/road test and check for correct operation.

Job card

Technician/learner name & date	Make and model	VIN no.	Reg. no.	Job/task no.

Customer's instructions/vehicle fault	Mileage	

Work carried out and recommendations (include PPE & special precautions taken)

Parts and labour	Price
Total	

Data and specifications used (include the actual figures)

Assessor report

	Assessment outcome	Passed (tick ✓)
1	The learner worked safely and minimised risks to themselves and others	
2	The learner correctly selected and used appropriate technical information	
3	The learner correctly selected and used appropriate tools and equipment	
4	The learner correctly carried out the task required using suitable methods and testing procedures	
5	The learner correctly recorded information and made suitable recommendations	

	Tick	Written feedback (with reference to assessment criteria) must be given when a learner is referred
Pass: I confirm that the learner's work was to an acceptable standard and met the assessment criteria of the unit		
Refer: The work carried out did not achieve the standards specified by the assessment criteria		

Assessor name (print)	Assessor PIN/ref.	Date

Section below only to be completed by the learner once the assessor decision has been made and feedback given			
I confirm that the work carried out was my own, and that I received feedback from the Assessor	Learner name (print)	Learner signature	Date

Worksheet 24: Remove and replace coil, primary and secondary circuit components

Procedure

▶ Disconnect battery ground lead. Identify type of ignition system.

▶ Label and disconnect cables to coil, primary and secondary circuit components. Undo and remove coil securing screws and remove coil. Inspect coil insulation and terminals for condition. When replacing a coil always check that the correct type is to be fitted. Check polarity, volts and ohms.

▶ Identify primary circuit cables and components – ballast resistor, starter solenoid leads, cables, switches, etc. Replace as per the manufacturer's instructions.

▶ Identify secondary circuit wires and components, pull out terminal connectors and undo securing screws or clips and remove components, distributor cap, rotor, plug wires and plugs.

▶ Inspect and test components. Replace in reverse order, reconnect battery ground lead, run engine and check for correct operation.

Job card

Technician/learner name & date	Make and model	VIN no.		Reg. no.	Job/task no.

Customer's instructions/vehicle fault		Mileage	

Work carried out and recommendations (include PPE & special precautions taken)

Parts and labour	Price
Total	

Data and specifications used (include the actual figures)

Assessor report

	Assessment outcome	Passed (tick ✓)
1	The learner worked safely and minimised risks to themselves and others	
2	The learner correctly selected and used appropriate technical information	
3	The learner correctly selected and used appropriate tools and equipment	
4	The learner correctly carried out the task required using suitable methods and testing procedures	
5	The learner correctly recorded information and made suitable recommendations	

	Tick	Written feedback (with reference to assessment criteria) must be given when a learner is referred
Pass: I confirm that the learner's work was to an acceptable standard and met the assessment criteria of the unit		
Refer: The work carried out did not achieve the standards specified by the assessment criteria		

Assessor name (print)	Assessor PIN/ref.	Date

Section below only to be completed by the learner once the assessor decision has been made and feedback given			
I confirm that the work carried out was my own, and that I received feedback from the Assessor	Learner name (print)	Learner signature	Date

Worksheet 25: Replace and/or clean fuel filters

Procedure

▶ For gasoline fuel systems, disconnect the battery earth lead. Depressurize fuel system if necessary. Apply vacuum to pressure regulator or cover with rag and undo a fuel supply union.

▶ For gasoline filter elements in fuel pumps and carburettor inlets, remove the cover, catch spilt fuel, clean out sediment chamber and water trap and clean with airline, clean filter with airline. Check seal and replace if necessary. Refit filter and cover, and tighten securing screws.

▶ For inline filters, either disconnect the fuel line above the tank or pinch a flexible hose with a brake pipe clamp before removing the filter in order to prevent fuel draining. Remove the pipes or hoses on each end of the filter, fit new filter observing the direction of flow arrow on the filter.

▶ For canister type filters, undo the canister body with a strap wrench or by hand. Drain canister from drain valve into a suitable small drain tray. Undo valve and unscrew the filter body to provide an air gap. Allow to drain before doing up the drain valve prior to removal.

▶ For replaceable filters in a canister bowl, undo the securing screw and lower the bowl, catch the spilt fuel in a drain tray.

▶ Clean out the bowl – check that the spring and plate are in position – replace the sealing rings and filter, fit in position and loosely tighten securing screw. Twist the bowl to seat the sealing rings and then finally tighten the securing screw.

▶ Run engine and check for correct operation.

Job card

Technician/learner name & date	Make and model	VIN no.	Reg. no.	Job/task no.

Customer's instructions/vehicle fault	Mileage	

Work carried out and recommendations (include PPE & special precautions taken)

Parts and labour	Price
Total	

Data and specifications used (include the actual figures)

Assessor report

	Assessment outcome	Passed (tick ✓)
1	The learner worked safely and minimised risks to themselves and others	
2	The learner correctly selected and used appropriate technical information	
3	The learner correctly selected and used appropriate tools and equipment	
4	The learner correctly carried out the task required using suitable methods and testing procedures	
5	The learner correctly recorded information and made suitable recommendations	

	Tick	Written feedback (with reference to assessment criteria) must be given when a learner is referred
Pass: I confirm that the learner's work was to an acceptable standard and met the assessment criteria of the unit		
Refer: The work carried out did not achieve the standards specified by the assessment criteria		

Assessor name (print)	Assessor PIN/ref.	Date

Section below only to be completed by the learner once the assessor decision has been made and feedback given			
I confirm that the work carried out was my own, and that I received feedback from the Assessor	Learner name (print)	Learner signature	Date

Worksheet 26: Remove and replace multipoint injection components

Procedure

▶ Disconnect the battery ground lead. Ensure a very clean work environment. Clean parts before removal.

▶ The fuel supply components are an electric rotary fuel pump fitted inside the fuel tank, an inline filter, the fuel pressure regulator and the injector valves on the fuel rail over the inlet manifold, and the fuel supply and return pipes and hoses.

▶ Identify the location of fuel supply components and disconnect and cap or plug fuel pipes/hoses to prevent loss of fuel and to keep clean. Undo securing devices/clamps/screws and undo union nuts on pipes to remove components. Inspect.

▶ The fuel rail, pressure regulator and injectors can be removed as a single unit and can be stripped for cleaning and replacement of parts. Follow manufacturer's instructions.

▶ Air supply components are the air box and air cleaner housing and filter element, ducting, the air temperature sensor and control flap, the air flow sensor, the throttle body assembly and the inlet manifold.

▶ Identify the location of the air supply components and disconnect ducting, securing screws and brackets, electrical terminal blocks, etc. and remove components. Inspect.

▶ The electronic components are the control relay in the main fuse box, the electronic control unit (ECU/ECM), sensors for air charge temperature, mass air flow meter, engine temperature, throttle valve switch for throttle positions, exhaust oxygen/lambda, and others (see manuals).

▶ The actuators include the injectors, auxiliary air valve or idle speed control valve, and the control devices for emission control systems.

▶ Identify the location of the electronic components and remove and replace in accordance with the manufacturer's instructions. Inspect components.

▶ After reassembly run engine, check for fuel leaks and carry out basic idle and mixture (gas analysis) adjustments.

Job card

Technician/learner name & date	Make and model	VIN no.		Reg. no.	Job/task no.

Customer's instructions/vehicle fault	Mileage	

Work carried out and recommendations (include PPE & special precautions taken)

Parts and labour	Price
Total	

Data and specifications used (include the actual figures)

Assessor report

	Assessment outcome	Passed (tick ✓)
1	The learner worked safely and minimised risks to themselves and others	
2	The learner correctly selected and used appropriate technical information	
3	The learner correctly selected and used appropriate tools and equipment	
4	The learner correctly carried out the task required using suitable methods and testing procedures	
5	The learner correctly recorded information and made suitable recommendations	

	Tick	Written feedback (with reference to assessment criteria) must be given when a learner is referred
Pass: I confirm that the learner's work was to an acceptable standard and met the assessment criteria of the unit		
Refer: The work carried out did not achieve the standards specified by the assessment criteria		

Assessor name (print)	Assessor PIN/ref.	Date

Section below only to be completed by the learner once the assessor decision has been made and feedback given			
I confirm that the work carried out was my own, and that I received feedback from the Assessor	Learner name (print)	Learner signature	Date

Worksheet 27: Remove and replace electronic sensors and ECUs

Procedure

▶ Follow the manufacturer's technical data for the correct procedures to identify remove and replace ignition system and associated systems sensors.

▶ Many sensors are held in place with one or two screws/bolts. Others are threaded and screw directly into the engine, air intake duct, exhaust manifold, etc. Disconnect cable termination blocks, undo and remove sensors.

▶ Locate ECU and disconnect main cable terminal block. Undo and remove securing screws and remove ECU. Inspect terminal sockets and pins.

▶ Refit in reverse order. Use sealant on sensor threads if specified. Check for the correct fitting of spacers, alignments, air gaps and positions as per manufacturer's specifications. Take care when reconnecting to protect terminal sockets and pins from damage.

▶ Check correct operation of sensors using a graphing multimeter (GMM), digital storage oscilloscope (DSO) or scan tool. Compare results to manufacturer's specifications.

Job card

Technician/learner name & date	Make and model	VIN no.		Reg. no.	Job/task no.

Customer's instructions/vehicle fault	Mileage	

Work carried out and recommendations (include PPE & special precautions taken)

Parts and labour	Price
Total	

Data and specifications used (include the actual figures)

Assessor report

	Assessment outcome	Passed (tick ✓)
1	The learner worked safely and minimised risks to themselves and others	
2	The learner correctly selected and used appropriate technical information	
3	The learner correctly selected and used appropriate tools and equipment	
4	The learner correctly carried out the task required using suitable methods and testing procedures	
5	The learner correctly recorded information and made suitable recommendations	

	Tick	Written feedback (with reference to assessment criteria) must be given when a learner is referred
Pass: I confirm that the learner's work was to an acceptable standard and met the assessment criteria of the unit		
Refer: The work carried out did not achieve the standards specified by the assessment criteria		

Assessor name (print)	Assessor PIN/ref.	Date

Section below only to be completed by the learner once the assessor decision has been made and feedback given			
I confirm that the work carried out was my own, and that I received feedback from the Assessor	Learner name (print)	Learner signature	Date

Worksheet 28: Remove and replace fuel injection air supply components

Procedure

▶ Disconnect the battery ground/earth lead. Identify the type of fuel injection system and depressurize in accordance with the manufacturer's instructions. Apply a vacuum to the pressure regulator.

▶ Undo the securing screws/clips on the air intake ducting, air cleaner and resonators. Remove components and inspect for condition, cleanliness and sealing at the joints.

▶ Label and disconnect the vacuum pipes/hoses and the electrical terminal blocks to the air flow meter, throttle body, auxiliary air valve, cold start valve and fuel injectors. Undo and remove the throttle cable and cruise control linkages.

▶ Undo and remove the securing screws for the air flow meter and the throttle body and the ducting in between. Remove components and inspect.

▶ Undo or unclip the securing devices for the fuel rail/pipes and fuel injectors and remove. Inspect sealing 'O' rings. Check, clean and test injectors before replacement.

▶ Undo and remove the plenum chamber and inlet manifold securing bolts and pull off components, inspect for condition and flatness of gasket faces and sealing ring grooves and matching bores.

▶ Reassemble in reverse order using new gaskets and sealing rings – lubricate gaskets and seals with lithium-based grease to aid fitting and sealing. Check on reassembly for correct positioning and sealing of components.

▶ Reconnect vacuum pipes/hoses, cables and electrical terminal blocks. Check throttle cable adjustment for specified free play at idle and throttle fully opened at the wide open throttle position.

▶ Run engine and check for correct operation. Listen for uneven running, flat spots and hesitation. Carry out exhaust gas analysis and adjust when necessary.

Job card

Technician/learner name & date	Make and model	VIN no.		Reg. no.	Job/task no.

Customer's instructions/vehicle fault	Mileage	

Work carried out and recommendations (include PPE & special precautions taken)

Parts and labour		Price
Total		

Data and specifications used (include the actual figures)

Assessor report

	Assessment outcome	Passed (tick ✓)
1	The learner worked safely and minimised risks to themselves and others	
2	The learner correctly selected and used appropriate technical information	
3	The learner correctly selected and used appropriate tools and equipment	
4	The learner correctly carried out the task required using suitable methods and testing procedures	
5	The learner correctly recorded information and made suitable recommendations	

	Tick	Written feedback (with reference to assessment criteria) must be given when a learner is referred
Pass: I confirm that the learner's work was to an acceptable standard and met the assessment criteria of the unit		
Refer: The work carried out did not achieve the standards specified by the assessment criteria		

Assessor name (print)	Assessor PIN/ref.	Date

Section below only to be completed by the learner once the assessor decision has been made and feedback given			
I confirm that the work carried out was my own, and that I received feedback from the Assessor	Learner name (print)	Learner signature	Date

Worksheet 29: Inspect fuel system for leaks and condition of pipes and hoses, etc.

Procedure

▶ This is a visual inspection and check for fuel odour under the vehicle and in the engine compartment.

▶ Raise the vehicle onto axle stands or a vehicle hoist. Open the bonnet. Follow the fuel lines from the tank to the carburettor of fuel injection components. Look carefully at all pipes and hoses. Look from the filler neck to the tank, at the tank, and the feed and return pipes.

▶ Look for washed areas, stains and fuel odour and any other signs of leakage. Check for damage, routing and chaffing. Look for corrosion, perishing, leaks at joints and security of hose clips and pipe securing clips.

▶ Look at the fuel tank security, and for leaks from the fuel gauge sender unit gasket and outlet pipe.

▶ Inspect the vapour lines, vapour trap and the filler cap fit and condition of the sealing ring. Modern tanks hold a small pressure or vacuum under some conditions. There is often a rush of air as the filler cap is removed – this is normal.

▶ Report any defects found.

Job card

Technician/learner name & date	Make and model	VIN no.		Reg. no.	Job/task no.
Customer's instructions/vehicle fault		Mileage			

Work carried out and recommendations (include PPE & special precautions taken)

Parts and labour	Price
Total	

Data and specifications used (include the actual figures)

Assessor report

	Assessment outcome	Passed (tick ✓)
1	The learner worked safely and minimised risks to themselves and others	
2	The learner correctly selected and used appropriate technical information	
3	The learner correctly selected and used appropriate tools and equipment	
4	The learner correctly carried out the task required using suitable methods and testing procedures	
5	The learner correctly recorded information and made suitable recommendations	

	Tick	Written feedback (with reference to assessment criteria) must be given when a learner is referred
Pass: I confirm that the learner's work was to an acceptable standard and met the assessment criteria of the unit		
Refer: The work carried out did not achieve the standards specified by the assessment criteria		

Assessor name (print)	Assessor PIN/ref.	Date

Section below only to be completed by the learner once the assessor decision has been made and feedback given			
I confirm that the work carried out was my own, and that I received feedback from the Assessor	Learner name (print)	Learner signature	Date

Worksheet 30: Remove and replace all parts of catalyst exhaust system including lambda/heated exhaust gas oxygen (HEGO) sensor

Procedure

▶ This task is similar to the replacement of a non-catalyst system but with additional items of the sensors, catalytic converter and heat shields.

▶ The converter can usually be removed from the system by undoing the ball joint or flange joint bolts at each end and the mounting bracket bolts and then lowering the catalyst out of the system. Be careful to provide full support for the catalyst. Inspect for condition inside and out.

▶ To remove the lambda/HEGO sensors, disconnect the terminal block and use a suitable spanner or special tool (see manufacturer's data) to unscrew the sensor. Undo the securing screws for heat shields and lower them from their mounting brackets. Check for damage or corrosion to the heat shields and the supporting brackets.

▶ Refit heat shields in reverse order ensuring the correct clearance between the heat shield and the vehicle floor. Refit sensors with a new sealing washer/gasket if required and tighten to the specified torque. Reconnect the electrical terminal block.

▶ Refit the catalytic converter using new gaskets or sealing compound (see manufacturer's data). Refit the mounting bracket and hangers and check the air gap between the converter and the heat shield and the heat shield and body (*very important*).

▶ Refit and align all exhaust system with new clamps and mountings/hangers and tighten clamp bolts. Check that the exhaust is clear of the body, steering and suspension components and brake pipes and any other components. Check for suspension travel but judging lowest suspension position.

▶ Run the engine and check for leaks. Carry out an exhaust gas analysis to ensure correct operation of the lambda/HEGO sensors and the catalytic converter.

Job card

Technician/learner name & date	Make and model	VIN no.		Reg. no.	Job/task no.

Customer's instructions/vehicle fault		Mileage	

Work carried out and recommendations (include PPE & special precautions taken)

Parts and labour	Price
Total	

Data and specifications used (include the actual figures)

Assessor report

	Assessment outcome	Passed (tick ✓)
1	The learner worked safely and minimised risks to themselves and others	
2	The learner correctly selected and used appropriate technical information	
3	The learner correctly selected and used appropriate tools and equipment	
4	The learner correctly carried out the task required using suitable methods and testing procedures	
5	The learner correctly recorded information and made suitable recommendations	

	Tick	Written feedback (with reference to assessment criteria) must be given when a learner is referred
Pass: I confirm that the learner's work was to an acceptable standard and met the assessment criteria of the unit		
Refer: The work carried out did not achieve the standards specified by the assessment criteria		

Assessor name (print)	Assessor PIN/ref.	Date

Section below only to be completed by the learner once the assessor decision has been made and feedback given			
I confirm that the work carried out was my own, and that I received feedback from the Assessor	Learner name (print)	Learner signature	Date

Worksheet 31: Check operation of positive crankcase ventilation system

Procedure

▶ Obtain driver information on oil and fuel consumption. Both are likely to be high if the system is blocked.

▶ Run engine and look for blue exhaust smoke and smoke leakage from PCV pipes and hoses. Exhaust smoke and engine smoke can indicate a blocked PCV system. Inspect the air cleaner element for oil contamination. If contaminated check oil separator.

▶ Identify PCV system type (valve or restricted orifice into inlet manifold). Visually inspect all pipes and hoses for condition, blockage, fit onto connectors (leakage), and oil contamination – pull off pipes in turn to assess internal and external condition.

▶ Clean or replace pipes/hoses if required. Clean or replace PCV air intake filter (integral with oil filler cap on some engines).

▶ Pull out and open PCV valve, clean and check operation or remove valve and shake to check that the valve is free, listen for a rattle. Check that air flow is toward manifold or air cleaner. *Or*, check restricted orifice into carburettor or inlet manifold is clean and clear, blow through and listen. Clean out if necessary (do not increase diameter by drilling clear with oversize drill).

▶ Reconnect all pipes and run engine. Check for vacuum at feed into inlet manifold. Check for symptoms of blocked valve or orifice – rough or uneven idle, stalling at low speeds, oil leaks, and smoke from engine oil filler cap.

Job card

Technician/learner name & date	Make and model	VIN no.		Reg. no.	Job/task no.
Customer's instructions/vehicle fault		Mileage			

Work carried out and recommendations (include PPE & special precautions taken)

Parts and labour	Price
Total	

Data and specifications used (include the actual figures)

Assessor report

	Assessment outcome	Passed (tick ✓)
1	The learner worked safely and minimised risks to themselves and others	
2	The learner correctly selected and used appropriate technical information	
3	The learner correctly selected and used appropriate tools and equipment	
4	The learner correctly carried out the task required using suitable methods and testing procedures	
5	The learner correctly recorded information and made suitable recommendations	

	Tick	Written feedback (with reference to assessment criteria) must be given when a learner is referred
Pass: I confirm that the learner's work was to an acceptable standard and met the assessment criteria of the unit		
Refer: The work carried out did not achieve the standards specified by the assessment criteria		

Assessor name (print)	Assessor PIN/ref.	Date

Section below only to be completed by the learner once the assessor decision has been made and feedback given			
I confirm that the work carried out was my own, and that I received feedback from the Assessor	Learner name (print)	Learner signature	Date

Worksheet 32: Check exhaust system for condition, leaks, blockage and security

Procedure

▶ Open bonnet – run engine and accelerate engine and listen for abnormal noises – exhaust blow, squeal, screech, excessive noise level and that the engine revs freely (possible blockage).

▶ Allow the engine to idle – cover the tail pipe outlet with a cloth pad – listen for exhaust blow.

▶ Inspect under the bonnet for exhaust manifold gasket leaks – look for tell-tale black markings from a loose or broken joint.

▶ Under the vehicle follow the exhaust pipes along – look at the down/front pipe to exhaust manifold joint and all the pipe to pipe or pipe to silencer/muffler/catalyst/resonator/etc. joints. Look for tell-tale black markings from loose, broken or badly sealed joints.

▶ Look closely at the skin of all pipes, silencers, mufflers, catalysts, resonators, etc. Look for corrosion, holes or other damage or deterioration.

▶ Look closely at the fitting, security, condition and positioning of heat shields. These must be in place to protect flammable materials inside the vehicle (sound deadening, etc.) from ignition from the high temperature of exhausts and catalytic converters.

▶ Look closely at all exhaust mounting brackets and rubber mounting components (hangers) for fitting, corrosion, perishing, separation, tension, etc. On turbocharged engines check the down/front pipe support bracket for condition and security.

▶ Check the position and routing of the exhaust for knocking on the body or chassis and for fouling on other components such as brake pipes, steering, suspension, axle, etc.

▶ If blockage is suspected, exhaust gas leaks in front of the blockage are likely. Disconnect a pipe at a suitable position in front of the blockage and check the flow through the remainder of the system with an air line blower – plug the pipe around the blower with a cloth pad.

Job card

Technician/learner name & date	Make and model	VIN no.		Reg. no.	Job/task no.
Customer's instructions/vehicle fault		Mileage			
Work carried out and recommendations (include PPE & special precautions taken)					
Parts and labour					Price
Total					
Data and specifications used (include the actual figures)					

Assessor report

Assessment outcome			Passed (tick ✓)
1	The learner worked safely and minimised risks to themselves and others		
2	The learner correctly selected and used appropriate technical information		
3	The learner correctly selected and used appropriate tools and equipment		
4	The learner correctly carried out the task required using suitable methods and testing procedures		
5	The learner correctly recorded information and made suitable recommendations		

	Tick	Written feedback (with reference to assessment criteria) must be given when a learner is referred
Pass: I confirm that the learner's work was to an acceptable standard and met the assessment criteria of the unit		
Refer: The work carried out did not achieve the standards specified by the assessment criteria		

Assessor name (print)	Assessor PIN/ref.	Date

Section below only to be completed by the learner once the assessor decision has been made and feedback given			
I confirm that the work carried out was my own, and that I received feedback from the Assessor	Learner name (print)	Learner signature	Date

Worksheet 33: Remove and replace supercharger

Procedure

▶ Disconnect battery ground lead. Disconnect and remove air ducting to the supercharger.

▶ Depending on the type of supercharger, identify aligning marks on the pulley and/or gears and the direction of rotation of the drive belt. Align timing marks if necessary, mark direction of rotation on the drive belt, slacken the drive belt tensioner and remove the belt.

▶ Undo and remove oil feed and return pipes.

▶ Undo bolts/nuts securing the supercharger and remove. Lift the supercharger from the adapter or mounting plate on the air intake to the engine.

▶ Replace in reverse order using new gaskets and tighten bolts/nuts to the manufacturer's specified torque settings and in sequence (centre outwards). Refit and adjust the tension of the drive belt (observe direction of rotation).

▶ Run engine and check for correct operation of the engine and the supercharger.

Job card

Technician/learner name & date	Make and model	VIN no.		Reg. no.	Job/task no.

Customer's instructions/vehicle fault		Mileage	

Work carried out and recommendations (include PPE & special precautions taken)

Parts and labour		Price
Total		

Data and specifications used (include the actual figures)

Assessor report

	Assessment outcome	Passed (tick ✓)
1	The learner worked safely and minimised risks to themselves and others	
2	The learner correctly selected and used appropriate technical information	
3	The learner correctly selected and used appropriate tools and equipment	
4	The learner correctly carried out the task required using suitable methods and testing procedures	
5	The learner correctly recorded information and made suitable recommendations	

	Tick	Written feedback (with reference to assessment criteria) must be given when a learner is referred
Pass: I confirm that the learner's work was to an acceptable standard and met the assessment criteria of the unit		
Refer: The work carried out did not achieve the standards specified by the assessment criteria		

Assessor name (print)	Assessor PIN/ref.	Date

Section below only to be completed by the learner once the assessor decision has been made and feedback given			
I confirm that the work carried out was my own, and that I received feedback from the Assessor	Learner name (print)	Learner signature	Date

Worksheet 34: Remove and replace exhaust gas recirculation (EGR) components

Procedure

▶ Disconnect battery ground lead. Identify the type and location of the EGR components from a vehicle workshop manual.

▶ For vacuum-controlled EGR systems, label and disconnect the vacuum and electrical connections at the vacuum control solenoid valve. Undo the securing screws and remove the solenoid valve.

▶ Label and remove the vacuum and electrical terminals from the EGR valve and EGR valve position sensor (integral with the EGR valve on closed-loop systems).

▶ Undo and remove the exhaust outlet pipe to the EGR valve and the feed pipe to the inlet manifold. Undo the EGR valve and remove.

▶ Reassemble in reverse order using new gaskets and seals as necessary.

▶ For electronic EGR valves, disconnect the electrical terminal block and the exhaust outlet and inlet manifold feed pipes. Remove the securing device and remove the valve (there are no separate control solenoid or vacuum pipes).

▶ Run engine and check for correct operation of the engine and EGR system.

Job card

Technician/learner name & date	Make and model	VIN no.		Reg. no.	Job/task no.

Customer's instructions/vehicle fault	Mileage	

Work carried out and recommendations (include PPE & special precautions taken)

Parts and labour	Price
Total	

Data and specifications used (include the actual figures)

Assessor report

	Assessment outcome	Passed (tick ✓)
1	The learner worked safely and minimised risks to themselves and others	
2	The learner correctly selected and used appropriate technical information	
3	The learner correctly selected and used appropriate tools and equipment	
4	The learner correctly carried out the task required using suitable methods and testing procedures	
5	The learner correctly recorded information and made suitable recommendations	

	Tick	Written feedback (with reference to assessment criteria) must be given when a learner is referred
Pass: I confirm that the learner's work was to an acceptable standard and met the assessment criteria of the unit		
Refer: The work carried out did not achieve the standards specified by the assessment criteria		

Assessor name (print)	Assessor PIN/ref.	Date

Section below only to be completed by the learner once the assessor decision has been made and feedback given			
I confirm that the work carried out was my own, and that I received feedback from the Assessor	Learner name (print)	Learner signature	Date

Worksheet 35: Remove and replace battery, battery cables and securing devices

Procedure

▶ Open bonnet/bonnet to access the battery. Check the general condition on and around the battery. Fit memory keeper (if available). Disconnect the earth or ground lead and then the supply lead from the battery.

▶ Undo and remove the battery retaining straps, clamps and brackets and remove. Remove any obstructions that may prevent the battery being lifted out. Plan the removal so that the lift, route and new position of the battery are known beforehand. Keeping the battery as level as possible lift up from the carrier and lift out from the vehicle. Place the battery on a bench or other suitable place.

▶ If it is necessary to use a lifting tool on the battery, ensure that it is correctly located on the battery casing, is secure and will not crush or otherwise damage the battery.

▶ Be very careful not to spill battery acid. If acid is split or has been lost from the battery, treat the area with a solution of baking soda and water or ammonia and water (an alkaline to neutralize the acid). Treat bare metal and repaint if necessary. Remove contaminated clothing and rinse as quickly as possible. Rinse acid from skin immediately. If acid burns are experienced, seek medical attention.

▶ Remove all cables by undoing securing bolts/nuts or pulling apart at the terminal or terminal block. Clean all terminals and connections to ensure good current flow.

▶ Reconnect and check the tightness of all push fit terminals. Coat with petroleum jelly or water repellent.

▶ Refit the battery, and if fitting a replacement, ensure that the new battery matches the old battery for casing dimensions, Ah capacity, and the type and position of the terminal posts. Fit retaining straps/clamps and tighten. Do not over tighten as battery damage can occur.

▶ Refit the battery supply lead and the earth or ground lead. Coat the terminals with petroleum jelly.

▶ Start and run the engine and check that the engine starter motor operated correctly and that the generator/ignition warning light came on and went out as the engine speed increased. Restore electronic memory functions if memory keeper not available.

Job card

Technician/learner name & date	Make and model	VIN no.		Reg. no.	Job/task no.

Customer's instructions/vehicle fault		Mileage	

Work carried out and recommendations (include PPE & special precautions taken)

Parts and labour	Price
Total	

Data and specifications used (include the actual figures)

Assessor report

	Assessment outcome	Passed (tick ✓)
1	The learner worked safely and minimised risks to themselves and others	
2	The learner correctly selected and used appropriate technical information	
3	The learner correctly selected and used appropriate tools and equipment	
4	The learner correctly carried out the task required using suitable methods and testing procedures	
5	The learner correctly recorded information and made suitable recommendations	

	Tick	Written feedback (with reference to assessment criteria) must be given when a learner is referred
Pass: I confirm that the learner's work was to an acceptable standard and met the assessment criteria of the unit		
Refer: The work carried out did not achieve the standards specified by the assessment criteria		

Assessor name (print)	Assessor PIN/ref.	Date

Section below only to be completed by the learner once the assessor decision has been made and feedback given			
I confirm that the work carried out was my own, and that I received feedback from the Assessor	Learner name (print)	Learner signature	Date

Worksheet 36: Inspect batteries for condition, security and state of charge

Procedure

▶ Open bonnet to access the battery. Check the general condition on and around the battery. Check the alternator drive belt for condition and tension. Look at the battery for signs of leakage from the casing. A build-up of a light coloured corrosive substance indicates the presence of battery acid. Look at the securing straps and brackets for security and condition. Look at the battery cables for condition, security and corrosion. Look at the earth or ground lead connection to the vehicle body for condition, security and corrosion. Make a general check of all cables to the starter and alternator and the engine earth or ground lead.

▶ Check the electrolyte level and specific gravity (relative density) with a hydrometer. *Or* check the built-in hydrometer on maintenance-free batteries. Green dot for charged, black for partially charged and yellow for internal faults where the battery should not be recharged. Carry out battery capacity test and confirm it has a capacity recommended by the vehicle manufacturer. Check the voltage with a digital voltmeter. Compare with the manufacturer's specifications. Conventional lead acid 12 volt batteries should read 13 to 13.2 volts when fully charged. Maintenance-free 12 volt batteries should read 12.8 volts when fully charged.

▶ Carry out a high rate discharge test *only* on a fully charged battery. Connect the tester following the manufacturer's instructions. Observe polarity. Set amperage to 3x battery Ah rate. Carry out test following the equipment manufacturer's instructions. *Do not* exceed test time. Conventional batteries can be 'fast charged' using a high output (amps) fast charger. Connect and charge at the current (amps) and time specified by the fast charger manufacturer. Disconnect the battery leads before connecting to the fast charger.

▶ *Do not* allow the battery to overheat. Use the temperature probe if fitted to the charger. *Do not* fast charge maintenance-free batteries unless permitted by the manufacturer. Slow (low current – amps) charging of conventional lead acid and maintenance-free batteries can be carried out singularly or two or more of a similar type and at a similar state of charge can be connected together.

▶ Connect a single battery to a charger observing the polarity of the charger and the battery. Red (+) to positive, black (-) to negative. Set the charger to 1/10th of the battery Ah (amp hour – 10 hour rating) and charge for 15 hours for a fully discharged battery. For multiple battery charging, connect batteries in parallel for a 12 volt charger. For 24 volt chargers, connect two 12 volt batteries in series. Four batteries are connected in series and parallel. Set the charge rate according to the number and Ah ratings of the batteries.

Job card

Technician/learner name & date	Make and model	VIN no.		Reg. no.	Job/task no.

Customer's instructions/vehicle fault	Mileage	

Work carried out and recommendations (include PPE & special precautions taken)

Parts and labour	Price
Total	

Data and specifications used (include the actual figures)

Assessor report

	Assessment outcome	Passed (tick ✓)
1	The learner worked safely and minimised risks to themselves and others	
2	The learner correctly selected and used appropriate technical information	
3	The learner correctly selected and used appropriate tools and equipment	
4	The learner correctly carried out the task required using suitable methods and testing procedures	
5	The learner correctly recorded information and made suitable recommendations	

	Tick	Written feedback (with reference to assessment criteria) must be given when a learner is referred
Pass: I confirm that the learner's work was to an acceptable standard and met the assessment criteria of the unit		
Refer: The work carried out did not achieve the standards specified by the assessment criteria		

Assessor name (print)	Assessor PIN/ref.	Date

Section below only to be completed by the learner once the assessor decision has been made and feedback given			
I confirm that the work carried out was my own, and that I received feedback from the Assessor	Learner name (print)	Learner signature	Date

Worksheet 37: Remove and replace alternator strip and replace brushes, regulator and diode pack

Procedure

▶ Disconnect the battery earth or ground lead. Unplug the multi-socket on the rear of the alternator or label and disconnect the cables. Slacken the alternator securing bolts, including the drive belt adjuster strap bolts. Mark the direction of rotation on the belt. Slacken the drive belt tension and remove the belt. Inspect the drive belt and drive pulleys for signs of wear, damage and slipping (glazing on sides).

▶ Hold the alternator secure and remove the securing bolts. Lift out the alternator. Check the spindle bearings by rotating and listening and feeling for uneven and lumpy movement. Check the alternator casing for damage. To remove the pulley, hold the spindle with a key or wrench and a wrench on the pulley securing nut. If the spindle cannot be held, fit an old belt around the pulley and hold the belt in a vice and then turn the pulley securing nut with a wrench. Pull off the pulley. Use a puller if necessary. Lift out the woodruff key if the casing is to be separated. The brushes are usually integral with the regulator on modern alternators. Undo the securing screws on the rear of the alternator and lift out the regulator and brush assembly. Inspect the brushes for length and condition. Inspect the slip rings on the rotor. Remove the rear cover. Undo and remove the through bolts holding the casing halves. Carefully separate the casing making sure that the stator remains in the rear casing and attached to the diode pack. Pull out the rotor from inside the stator.

▶ Remove the circlips retaining the bearings, in each casing half and gently tap out the bearings. Check that the bearings have not been spinning in the casing. Fit new bearings and pack with grease. Fit new circlips and seals. Pre-packed and integral seal bearings are sometimes used. Where a diode pack is available as a replacement, part undo the retaining screws holding the diode pack to the rear casing and unsolder the old diodes where they attach to the stator. Re-solder the new diodes using a *heat sink* such as a pair of pliers to prevent heat damage to the diodes.

▶ Reassemble the alternator in the reverse order. Spin before refitting to check for abnormal noises. Carry out a bench test if the equipment is available. Refit in the reverse order and adjust the drive belt tension. Run the engine and check that the generator/ignition warning light comes on and then goes out as the engine speed increases. Connect a digital voltmeter and clamp on ammeter and check that the alternator output is correct. Compare with manufacturer's data.

Job card

Technician/learner name & date	Make and model	VIN no.		Reg. no.	Job/task no.

Customer's instructions/vehicle fault	Mileage	

Work carried out and recommendations (include PPE & special precautions taken)	

Parts and labour	Price
Total	

Data and specifications used (include the actual figures)

Assessor report

	Assessment outcome	Passed (tick ✓)
1	The learner worked safely and minimised risks to themselves and others	
2	The learner correctly selected and used appropriate technical information	
3	The learner correctly selected and used appropriate tools and equipment	
4	The learner correctly carried out the task required using suitable methods and testing procedures	
5	The learner correctly recorded information and made suitable recommendations	

	Tick	Written feedback (with reference to assessment criteria) must be given when a learner is referred
Pass: I confirm that the learner's work was to an acceptable standard and met the assessment criteria of the unit		
Refer: The work carried out did not achieve the standards specified by the assessment criteria		

Assessor name (print)	Assessor PIN/ref.	Date

Section below only to be completed by the learner once the assessor decision has been made and feedback given			
I confirm that the work carried out was my own, and that I received feedback from the Assessor	Learner name (print)	Learner signature	Date

Worksheet 38: Remove and replace alternator circuit cables/components

Procedure

▶ Disconnect the battery earth or ground lead.

▶ Apart from the alternator, the main components are the alternator output cable to the battery, the battery feed to the ignition switch, the ignition switch, generator warning light and the feed to the field terminal on the alternator.

▶ Follow the manufacturer's instructions for replacement of the ignition switch and the fascia for access to the warning light bulb and printed circuit board on the rear of the instrument panel. Inspect the printed circuit, bulb holder and instruments multi-socket termination.

▶ Label and remove cables.

▶ Reassemble in reverse order. Run the engine and check the operation of the alternator and charge circuit.

▶ If replacing an old type generator of a dynamo type, it is important that the field coils are polarized. Take a live feed with a jump lead from the battery feed terminal directly to the field coil terminal after reconnecting the battery leads but before connecting the field terminal. Hold in contact for 30 seconds.

Job card

Technician/learner name & date	Make and model	VIN no.		Reg. no.	Job/task no.

Customer's instructions/vehicle fault	Mileage	

Work carried out and recommendations (include PPE & special precautions taken)	

Parts and labour	Price
Total	

Data and specifications used (include the actual figures)

Assessor report

Assessment outcome		Passed (tick ✓)
1	The learner worked safely and minimised risks to themselves and others	
2	The learner correctly selected and used appropriate technical information	
3	The learner correctly selected and used appropriate tools and equipment	
4	The learner correctly carried out the task required using suitable methods and testing procedures	
5	The learner correctly recorded information and made suitable recommendations	

	Tick	Written feedback (with reference to assessment criteria) must be given when a learner is referred
Pass: I confirm that the learner's work was to an acceptable standard and met the assessment criteria of the unit		
Refer: The work carried out did not achieve the standards specified by the assessment criteria		

Assessor name (print)	Assessor PIN/ref.	Date

Section below only to be completed by the learner once the assessor decision has been made and feedback given			
I confirm that the work carried out was my own, and that I received feedback from the Assessor	Learner name (print)	Learner signature	Date

Worksheet 39: Remove and replace pre-engaged starter motor and replace solenoid

Procedure

▶ Check the condition of the battery and the operation of the starter motor before removal. Disconnect the battery earth or ground lead. Label and disconnect the cables on the solenoid – solenoid feed from the starter switch and LT ignition feed to bypass the ballast resistor. Disconnect the main supply feed cable to the starter at either end (most convenient). Remove any components that restrict removal of the starter motor. Undo the starter motor securing bolts. (Socket wrenches and extensions are useful to gain access between the starter motor body and the engine block). Carefully remove the starter motor. Inspect the drive pinion (gear) and the one-way clutch. Check the starter motor casing and fitting flange for damage. Check the starter ring gear on the flywheel through the hole where the starter motor fits. Turn the engine at least one full revolution.

▶ Check all gear teeth for chipped or worn down sections both on individual teeth and around the full circumference of the gear. Uneven wear in one place is a frequent defect. Look at the teeth for wear marks that show that the starter pinion has been fully in mesh. To strip the motor, fit into a bench vice and secure. To remove the solenoid, undo the feed cable into the motor and pull back out of the way. Undo the retaining screws in the casing. Carefully pull out the solenoid disconnecting the plunger from the pinion gear engagement lever. To strip the motor, remove the through bolts and the rear cover and brushes. The brushes may be available as replacement parts. Follow the manufacturer's instructions for replacement. Clean and inspect the commutator, armature and spindle bearings. Pull the pinion casing, pinion, engagement lever and armature from the main casing and field coils. Follow the manufacturer's instructions for replacing the drive pinion and one-way clutch.

▶ Reassemble in reverse order. Carry out a bench test with the starter securely held in a bench vice. Connect a battery with jump leads to the starter. Negative cable to the battery negative and the starter casing. Positive lead *only* to the battery positive and then keep the lead clear until testing. *Avoid* any connection of the positive lead to the starter casing, vice or bench. Test the solenoid operation by touching the lead to the solenoid terminal, which should click, and the pinion move along the spindle to the engaged position. The spindle may slowly revolve on some motors.

▶ Test the motor operation by touching the lead to the motor terminal. The motor should run at full speed. Connect the lead to the solenoid input terminal. The solenoid should not operate and the motor should not run. Use a jump lead to connect a feed to the solenoid low current terminal. The solenoid should operate and the motor should run. Refit the starter motor to the engine and check the operation.

Job card

Technician/learner name & date	Make and model	VIN no.		Reg. no.	Job/task no.

Customer's instructions/vehicle fault	Mileage	

Work carried out and recommendations (include PPE & special precautions taken)

Parts and labour	Price
Total	

Data and specifications used (include the actual figures)

Assessor report

	Assessment outcome	Passed (tick ✓)
1	The learner worked safely and minimised risks to themselves and others	
2	The learner correctly selected and used appropriate technical information	
3	The learner correctly selected and used appropriate tools and equipment	
4	The learner correctly carried out the task required using suitable methods and testing procedures	
5	The learner correctly recorded information and made suitable recommendations	

	Tick	Written feedback (with reference to assessment criteria) must be given when a learner is referred
Pass: I confirm that the learner's work was to an acceptable standard and met the assessment criteria of the unit		
Refer: The work carried out did not achieve the standards specified by the assessment criteria		

Assessor name (print)		Assessor PIN/ref.	Date

Section below only to be completed by the learner once the assessor decision has been made and feedback given			
I confirm that the work carried out was my own, and that I received feedback from the Assessor	Learner name (print)	Learner signature	Date

Chassis

Worksheet 40: Remove and refit front suspension strut and spring

Procedure

▶ *Removal.* Apply the handbrake, jack up the front of the car and support on stands. Remove the road wheel. To prevent the lower arm assembly hanging down whilst the strut is removed, screw a wheel bolt into the hub, then wrap a piece of wire around the bolt and tie it to the car body. This will support the weight of the hub assembly. Unclip the brake hose and wiring harness from any clips on the base of the strut. Slacken and remove the lower bolts securing the suspension strut to the steering knuckle. From within the engine compartment, unscrew the strut upper mounting nuts and then carefully lower the strut assembly out from underneath the wing.

▶ *Overhaul.* Warning: before dismantling the front suspension strut, a special tool to hold the coil spring in compression must be obtained. Any attempt to dismantle the strut without such a tool is likely to result in damage and/or injury. With the strut removed from the car, clean away all external dirt and then mount it upright in a vice. Fit the spring compressor, and compress the coil spring until all tension is relieved from the upper spring seat. Remove the cap from the top of the strut to gain access to the strut upper mounting retaining nut. Slacken the nut whilst holding the strut piston.

▶ Remove the mounting nut and washer, and lift off the rubber mounting plate. Remove the gasket and dished washer followed by the upper spring plate and upper spring seat. Lift off the coil spring and remove the lower spring seat. Examine all the components for wear, damage or deformation, and check the upper mounting bearing for smoothness of operation. Renew as necessary.

▶ Examine the strut for signs of fluid leakage. Check the strut piston for signs of pitting along its entire length, and check the strut body for signs of damage. While holding it in an upright position, test the operation of the strut by moving the piston through a full stroke, and then through short strokes (50 to 100 mm). In both cases, the resistance felt should be smooth and continuous. Renew if the resistance is jerky, uneven or if there is any visible sign of wear or damage. If any doubt exists about the condition of the coil spring, carefully remove the spring compressors, and check the spring for distortion and signs of cracking. Renew the spring if it is damaged.

▶ Fit the spring seat and coil spring onto the strut, making sure the spring end is correctly located against the strut stop. Fit the upper spring plate, aligning its stop with that of the seat, and fit the dished washer and gasket followed by the upper mounting plate. Locate the washer on the strut piston, then fit the mounting plate nut and tighten it to the specified torque. Refit all in reverse order.

Job card

Technician/learner name & date	Make and model	VIN no.	Reg. no.	Job/task no.

Customer's instructions/vehicle fault	Mileage	

Work carried out and recommendations (include PPE & special precautions taken)

Parts and labour	Price
Total	

Data and specifications used (include the actual figures)

Assessor report

	Assessment outcome	Passed (tick ✓)
1	The learner worked safely and minimised risks to themselves and others	
2	The learner correctly selected and used appropriate technical information	
3	The learner correctly selected and used appropriate tools and equipment	
4	The learner correctly carried out the task required using suitable methods and testing procedures	
5	The learner correctly recorded information and made suitable recommendations	

	Tick	Written feedback (with reference to assessment criteria) must be given when a learner is referred
Pass: I confirm that the learner's work was to an acceptable standard and met the assessment criteria of the unit		
Refer: The work carried out did not achieve the standards specified by the assessment criteria		

Assessor name (print)	Assessor PIN/ref.	Date

Section below only to be completed by the learner once the assessor decision has been made and feedback given			
I confirm that the work carried out was my own, and that I received feedback from the Assessor	Learner name (print)	Learner signature	Date

Worksheet 41: Remove and refit rear suspension spring

Procedure

Note: procedures vary so check the manufacturer's data before starting work.

▶ Removal:

- Chock the front wheels, jack up the rear of the car and support on stands. Remove the road wheel. If necessary, remove the bolts and plates securing the driveshaft to the final drive unit flange. Free the driveshaft and support it by tying it to the car under body using a piece of wire.
- Position a jack underneath the rear of the trailing arm, and support the weight of the arm.
- Slacken and remove the damper/shock absorber lower mounting bolt.
- Slowly lower the trailing arm. Keep an eye on the brake pipe/hose to ensure no excess strain is placed on it, until it is possible to withdraw the coil spring. Remove the spring seats.
- Inspect the spring closely for signs of damage, such as cracking, and check the spring seats for signs of wear.
- Renew worn components as necessary.

▶ Refitting:

- Fit the upper and lower spring seats, making sure they are correctly located.
- Engage the spring with its upper seat.
- Hold the spring in position and carefully raise the trailing arm whilst aligning the coil spring with its lower seat.
- Raise the arm fully and refit the damper/shock absorber lower mounting bolt, tightening it to the correct torque.
- If removed, connect the driveshaft to the final drive unit.
- Refit the road wheel then lower the car to the ground. Tighten the wheel bolts to the specified torque.

Job card

Technician/learner name & date	Make and model	VIN no.	Reg. no.	Job/task no.

Customer's instructions/vehicle fault	Mileage	

Work carried out and recommendations (include PPE & special precautions taken)

Parts and labour	Price
Total	

Data and specifications used (include the actual figures)

Assessor report

	Assessment outcome	Passed (tick ✓)
1	The learner worked safely and minimised risks to themselves and others	
2	The learner correctly selected and used appropriate technical information	
3	The learner correctly selected and used appropriate tools and equipment	
4	The learner correctly carried out the task required using suitable methods and testing procedures	
5	The learner correctly recorded information and made suitable recommendations	

	Tick	Written feedback (with reference to assessment criteria) must be given when a learner is referred
Pass: I confirm that the learner's work was to an acceptable standard and met the assessment criteria of the unit		
Refer: The work carried out did not achieve the standards specified by the assessment criteria		

Assessor name (print)	Assessor PIN/ref.	Date

Section below only to be completed by the learner once the assessor decision has been made and feedback given		
I confirm that the work carried out was my own, and that I received feedback from the Assessor	Learner name (print)	Learner signature	Date

Worksheet 42: Remove, inspect and replace lower suspension arm, bushes and ball joint

Procedure

▶ Remove:

- Disconnect the battery, apply the handbrake and slacken the road wheel nuts. Raise the front of the vehicle, support it on axle stands and remove the front wheels.
- Remove stabilizer and tie bar attachments to the lower arm. Using ball joint splitter, disconnect track rod end from steering arm. Place screw jack under hub assembly and slightly compress spring.
- Remove upper and lower shock absorber mountings and remove shock absorber.
- Fit spring compressor and compress coil spring. Using ball joint splitter, disconnect lower arm ball joint whilst supporting lower arm.
- Remove coil spring from vehicle and spring compressors from spring. Remove lower arm pivot bolt and lower arm from vehicle.

▶ Inspect:

- Inspect lower arm, spring insulator, bushing and ball joint – replace as appropriate.

Note: Replacement of bushes and ball joints is only possible on some vehicles. Refer to manufacturer's procedures if possible.

- Inspect coil spring.

▶ Refit:

- Refitting is a direct reversal of the above procedure.

Note: The spring must be compressed to approximately kerb weight position before refitting.

- The nuts on the lower pivot bolt must be tightened to the correct torque when the vehicle has been lowered onto its wheels. Check castor and wheel alignment. Reset if required.

Note: Wheel alignment must be measured and reset after castor has been adjusted.

Job card

Technician/learner name & date	Make and model	VIN no.		Reg. no.	Job/task no.

Customer's instructions/vehicle fault	Mileage	

Work carried out and recommendations (include PPE & special precautions taken)

Parts and labour	Price
Total	

Data and specifications used (include the actual figures)

Assessor report

	Assessment outcome	Passed (tick ✓)
1	The learner worked safely and minimised risks to themselves and others	
2	The learner correctly selected and used appropriate technical information	
3	The learner correctly selected and used appropriate tools and equipment	
4	The learner correctly carried out the task required using suitable methods and testing procedures	
5	The learner correctly recorded information and made suitable recommendations	

	Tick	Written feedback (with reference to assessment criteria) must be given when a learner is referred
Pass: I confirm that the learner's work was to an acceptable standard and met the assessment criteria of the unit		
Refer: The work carried out did not achieve the standards specified by the assessment criteria		

Assessor name (print)	Assessor PIN/ref.	Date

Section below only to be completed by the learner once the assessor decision has been made and feedback given			
I confirm that the work carried out was my own, and that I received feedback from the Assessor	Learner name (print)	Learner signature	Date

Worksheet 43: Remove, inspect and replace rear suspension control arm and bushes

Procedure

▶ Remove:

- Disconnect the battery, apply the handbrake and slacken the road wheel nuts. Raise the front of the vehicle, support it on axle stands and remove the front wheels. Remove stabilizer attachments to the lower arm. Clamp flexible brake pipe and disconnect from connector on shock absorber.
- Position jack under lower arm to take weight, remove tie bar to body retention bolt. Remove lower arm pivot bolt and lower shock absorber/strut connection. Slowly lower jack and remove lower arm and coil spring.

▶ Inspect:

- Inspect the lower arm and bushes, spring insulator and spring. Replace as necessary.

Note: Replacement of bushes is only possible on some vehicles. Refer to manufacturer's procedures if possible.

▶ Refit:

- Reposition lower arm, spring insulator and spring on jack and raise into position. Refit lower arm pivot bolt and lower shock absorber/strut connection.

Note: The lower pivot bolt must be tightened to the correct torque when the vehicle has been lowered onto its wheels.

- Refit tie bar to body retention mounting, tightening to the correct torque. Connect flexible brake pipe.
- Refit stabilizer attachments to the lower arm tightening to the correct torque.
- Bleed the brakes.
- Lower the vehicle to the ground. Tighten the lower pivot bolt to the correct torque. Tighten the wheel bolts to the correct torque.
- Check and adjust rear wheel camber, alignment and thrust angle.

Note: It is not possible to adjust rear wheel alignment on some vehicles. It is essential to fit all washers and spacers in exactly the same order as they were before the commencement of the repair operation.

Job card

Technician/learner name & date	Make and model	VIN no.	Reg. no.	Job/task no.

Customer's instructions/vehicle fault		Mileage	

Work carried out and recommendations (include PPE & special precautions taken)

	Price
Parts and labour	
Total	

Data and specifications used (include the actual figures)

Assessor report

	Assessment outcome	Passed (tick ✓)
1	The learner worked safely and minimised risks to themselves and others	
2	The learner correctly selected and used appropriate technical information	
3	The learner correctly selected and used appropriate tools and equipment	
4	The learner correctly carried out the task required using suitable methods and testing procedures	
5	The learner correctly recorded information and made suitable recommendations	

	Tick	Written feedback (with reference to assessment criteria) must be given when a learner is referred
Pass: I confirm that the learner's work was to an acceptable standard and met the assessment criteria of the unit		
Refer: The work carried out did not achieve the standards specified by the assessment criteria		

Assessor name (print)	Assessor PIN/ref.	Date

Section below only to be completed by the learner once the assessor decision has been made and feedback given			
I confirm that the work carried out was my own, and that I received feedback from the Assessor	Learner name (print)	Learner signature	Date

Worksheet 44: Remove, inspect and replace rear McPherson strut and spring

Procedure

▶ Remove:
  - Disconnect the battery, apply the handbrake and slacken the road wheel nuts. Raise the rear of the vehicle, support it on axle stands and remove the wheels.
  - Unclip the flexible brake hose and wiring harness from any clips on the base of the strut.
  - Position jack under lower arm. This will support the weight of the hub assembly.
  - From within the boot, unscrew the strut upper mounting nuts.
  - Slacken and remove the bolts securing the suspension strut to the hub assembly and then carefully lower the strut assembly out from underneath the wing.

▶ Inspect/overhaul:
  - Warning: before dismantling the front suspension strut, a special tool to hold the coil spring in compression must be obtained. Any attempt to dismantle the strut without such a tool is likely to result in damage and/or injury.
  - With the strut removed from the car, clean away all external dirt and then mount it upright in a vice. Fit the spring compressor, and compress the coil spring until all tension is relieved from the upper spring seat.
  - Remove the mounting nut and washer, and lift off the rubber mounting plate. Remove the gasket and dished washer followed by the upper spring plate and upper spring seat. Lift off the coil spring and remove the lower spring seat.
  - Examine all the components for wear, damage or deformation. Renew as necessary.
  - Examine the strut for signs of fluid leakage. Check the strut piston for signs of pitting along its entire length, and check the strut body for signs of damage. While holding it in an upright position, test the operation of the strut by moving the piston through a full stroke, and then through short strokes (50 to 100 mm). In both cases, the resistance felt should be smooth and continuous. Renew if the resistance is jerky, uneven or if there is any visible sign of wear or damage.
  - If any doubt exists about the condition of the coil spring, carefully remove the spring compressors, and check the spring for distortion and signs of cracking. Renew the spring if it is damaged or distorted, or if there is any doubt as to its condition. Inspect all other components for damage or deterioration, and renew any that are suspect.
  - Fit the spring seat and coil spring onto the strut, making sure the spring end is correctly located against the strut stop.
  - Fit the upper spring plate, aligning its stop with that of the seat, and fit the dished washer and gasket followed by the upper mounting plate. Locate the washer on the strut piston, then fit the mounting plate nut and tighten it to the specified torque.
  - Ensure the spring ends and seats are correctly located, then carefully release the compressor and remove it from the strut.

▶ Refitting:
  - Manoeuvre the strut assembly into position, and fit the upper mounting nuts.
  - Refit the bolts securing the suspension strut to the hub assembly.
  - Tighten the strut upper and lower mounting nuts to the specified torque.
  - Clip the hose/wiring back onto the strut, then refit the road wheel. Lower the car to the ground and tighten the wheel bolts to the correct torque.

Job card

Technician/learner name & date	Make and model	VIN no.		Reg. no.	Job/task no.

Customer's instructions/vehicle fault		Mileage		

Work carried out and recommendations (include PPE & special precautions taken)

Parts and labour	Price
Total	

Data and specifications used (include the actual figures)

Assessor report

	Assessment outcome	Passed (tick ✓)
1	The learner worked safely and minimised risks to themselves and others	
2	The learner correctly selected and used appropriate technical information	
3	The learner correctly selected and used appropriate tools and equipment	
4	The learner correctly carried out the task required using suitable methods and testing procedures	
5	The learner correctly recorded information and made suitable recommendations	

	Tick	Written feedback (with reference to assessment criteria) must be given when a learner is referred
Pass: I confirm that the learner's work was to an acceptable standard and met the assessment criteria of the unit		
Refer: The work carried out did not achieve the standards specified by the assessment criteria		

Assessor name (print)	Assessor PIN/ref.	Date

Section below only to be completed by the learner once the assessor decision has been made and feedback given			
I confirm that the work carried out was my own, and that I received feedback from the Assessor	Learner name (print)	Learner signature	Date

Worksheet 45: Remove and replace pitman arm, relay rod, idler arm and steering damper

Procedure

▶ Removal:

- Disconnect the battery, apply the handbrake and slacken the road wheel nuts. Raise the front of the vehicle, support it on axle stands and remove the front wheels.
- Disconnect the relay arm ball joints from the pitman arm and idler arm. Mark the relative position of the pitman arm to the rocker shaft and then remove the pitman arm using a suitable puller.
- Remove the idler arm and mountings. Remove steering linkage damper/shock absorber.

▶ Inspect:

- Inspect the steering box seals for leakage, the ball joints for play and the idler arm bushes for play.
- Adjust pinion worm bearing pre-load and sector lash.
- Examine the steering damper for signs of fluid leakage. Check the piston for signs of pitting along its entire length, and check the body for signs of damage. Test the operation of the damper/shock absorber by moving the piston through a full stroke, and then through short strokes. In both cases, the resistance felt should be smooth and continuous. Renew if the resistance is jerky, uneven or if there is any visible sign of wear or damage.

▶ Refit:

- Refitting is a direct reversal of the removal procedure. All fixings should be tightened to the manufacturer's recommended torque settings.
- Refit the road wheel. Lower the car to the ground and tighten the wheel bolts to the correct torque.
- Check wheel alignment. Reset if necessary.

Job card

Technician/learner name & date	Make and model	VIN no.	Reg. no.	Job/task no.

Customer's instructions/vehicle fault	Mileage	

Work carried out and recommendations (include PPE & special precautions taken)

Parts and labour	Price
Total	

Data and specifications used (include the actual figures)

Assessor report

	Assessment outcome	Passed (tick ✓)
1	The learner worked safely and minimised risks to themselves and others	
2	The learner correctly selected and used appropriate technical information	
3	The learner correctly selected and used appropriate tools and equipment	
4	The learner correctly carried out the task required using suitable methods and testing procedures	
5	The learner correctly recorded information and made suitable recommendations	

	Tick	Written feedback (with reference to assessment criteria) must be given when a learner is referred
Pass: I confirm that the learner's work was to an acceptable standard and met the assessment criteria of the unit		
Refer: The work carried out did not achieve the standards specified by the assessment criteria		

Assessor name (print)	Assessor PIN/ref.	Date

Section below only to be completed by the learner once the assessor decision has been made and feedback given			
I confirm that the work carried out was my own, and that I received feedback from the Assessor	Learner name (print)	Learner signature	Date

Worksheet 46: Remove, overhaul and refit steering rack

Procedure

▶ Remove:

- Disconnect the battery, apply the handbrake and slacken the front road wheel nuts. Raise the front of the vehicle, support it on stands and remove the front wheels. Disconnect the ball joints from the steering levers. Pull back the carpet and disconnect the intermediate shaft universal joint from the pinion shaft. Remove the steering rack mounting bolts, clamp plate and plastic seating. Manoeuvre the rack and pinion clear of the body.

- Withdraw the steering rack from the passenger's or driver's side of the engine compartment as appropriate. Remove the rack pinion cover seal.

▶ Overhaul:

- Remove both ball joints from the tie-rods and remove the seals and clips from the ends of the rack housing. Hold the rack in a soft-jawed vice and unscrew the ball housing from the rack.

- Tighten the ball housing onto the rack, to the correct torque and secure by staking the edge of the ball housing into the groove in the rack. Replenish any lubricant lost, and fit the rack seals and retaining clips. Screw the ball joint locknuts onto the tie-rods and screw each ball joint on an equal amount.

▶ Refit:

- Fit the steering rack through the passenger's or driver's side of the engine compartment. Position the pinion in the body aperture and loosely fit the bolts on the pinion side of the rack. Fit the plastic seating and clamp; fully tighten the clamp bolts, to the correct torque. Finally, tighten the bolts on the pinion side of the rack to the correct torque.

- Centralize the rack. Fit the pinion cover plate, position the steering wheel in the straight-ahead position and connect the universal joint to the pinion shaft. Fit the clamp bolt. Secure the sound deadening material around the pinion shaft and fit the carpet.

- Connect the steering ball pins to the steering levers. Fit the road wheels, lower the vehicle to the ground, and tighten the wheel nuts to the correct torque. Connect the battery. Check and, if necessary, adjust the front wheel alignment.

Job card

Technician/learner name & date	Make and model	VIN no.		Reg. no.	Job/task no.

Customer's instructions/vehicle fault		Mileage	

Work carried out and recommendations (include PPE & special precautions taken)

Parts and labour	Price
Total	

Data and specifications used (include the actual figures)

Assessor report

	Assessment outcome	Passed (tick ✓)
1	The learner worked safely and minimised risks to themselves and others	
2	The learner correctly selected and used appropriate technical information	
3	The learner correctly selected and used appropriate tools and equipment	
4	The learner correctly carried out the task required using suitable methods and testing procedures	
5	The learner correctly recorded information and made suitable recommendations	

	Tick	Written feedback (with reference to assessment criteria) must be given when a learner is referred
Pass: I confirm that the learner's work was to an acceptable standard and met the assessment criteria of the unit		
Refer: The work carried out did not achieve the standards specified by the assessment criteria		

Assessor name (print)	Assessor PIN/ref.	Date

Section below only to be completed by the learner once the assessor decision has been made and feedback given			
I confirm that the work carried out was my own, and that I received feedback from the Assessor	Learner name (print)	Learner signature	Date

Worksheet 47: Remove, inspect and refit PAS pump

Procedure

▶ Removal:

- Remove the cap from the fluid reservoir.

- Using suitable wrench, hold the adaptor and remove the high-pressure hose. Disconnect the low-pressure hose and drain the fluid into a container. Blank off the hoses and pump to prevent the ingress of dust and dirt. Replace the reservoir cap.

- Slacken and remove the adjustment bolts, situated underneath the pump. Remove the drive belt. Remove the two pivot bolts, withdraw the pump and bracket from the vehicle.

▶ Inspect:

- Check condition of drive belt, pump pulley, pump mounts, seals and gaskets, and hoses and fittings.

▶ Refit:

- Fit the pump assembly to the engine bracket, feeding the drive belt over the pulley, locate and secure the two pivot bolts but do not fully tighten. Fit the adjustment bolts. Check pulley alignment.

- Adjust the drive belt to the correct tension and tighten the pivot nuts and bolts.

- Remove the blanking plugs, connect the inlet hose to the adaptor and tighten to the correct torque.

- Fill the fluid reservoir to the 'MAX' mark, and fit the filler cap. Do not overfill.

- Check system operation and check for leaks.

Job card

Technician/learner name & date	Make and model	VIN no.	Reg. no.	Job/task no.

Customer's instructions/vehicle fault	Mileage	

Work carried out and recommendations (include PPE & special precautions taken)

Parts and labour	Price
Total	

Data and specifications used (include the actual figures)

Assessor report

	Assessment outcome	Passed (tick ✓)
1	The learner worked safely and minimised risks to themselves and others	
2	The learner correctly selected and used appropriate technical information	
3	The learner correctly selected and used appropriate tools and equipment	
4	The learner correctly carried out the task required using suitable methods and testing procedures	
5	The learner correctly recorded information and made suitable recommendations	

	Tick	Written feedback (with reference to assessment criteria) must be given when a learner is referred
Pass: I confirm that the learner's work was to an acceptable standard and met the assessment criteria of the unit		
Refer: The work carried out did not achieve the standards specified by the assessment criteria		

Assessor name (print)	Assessor PIN/ref.	Date

Section below only to be completed by the learner once the assessor decision has been made and feedback given			
I confirm that the work carried out was my own, and that I received feedback from the Assessor	Learner name (print)	Learner signature	Date

Worksheet 48: Remove and refit wheel bearings

Procedure

▶ *Front hub assembly.* Apply the handbrake and slacken the front road wheel nuts, raise the front of the vehicle, support it on stands and remove the road wheel. Remove the drive shaft nut split pin, use an assistant to apply firm pressure to the brake pedal and, while the brake is applied, unscrew the drive shaft nut. Remove the brake caliper and support it to prevent straining the hose. Remove the disc. Using a ball joint breaker tool, disconnect the ball joint from the steering lever. Unscrew the nuts and remove the bolts to release the strut from the hub assembly. Unscrew the nut and remove the clamp bolt, securing the lower ball joint to the hub assembly and, using a suitable lever placed between the lower arm and the anti-roll bar, lever downwards to release the ball joint from the hub. Remove the hub assembly from the drive shaft.

▶ Extract the inner oil seal and spacer and the outer oil seal. Drive out one of the bearings, invert the hub and drive out the remaining bearing. Inspect the bearings for signs of wear and damage, renew as necessary. Pack the bearings with suitable grease and press them into the hub. Fit the oil seal spacer against the inner bearing. Fit the oil seals. Locate the hub on the drive shaft, fit the flat washer and drive shaft nut and tighten the nut finger tight. Fit the hub assembly to the lower ball joint, fit the clamp bolt and tighten the nut. Fit the hub to the strut, fit the bolts and tighten the nuts to the correct torque. Connect the ball joint to the steering lever and fit and tighten the nut. Fit the disc to the drive flange and tighten the securing screws. Fit the brake caliper. Use an assistant to apply firm pressure to the brake pedal and, while the brake is applied, tighten the drive shaft nut to the correct torque. Lock the nut with a new split pin.

▶ *Rear hub assembly.* Chock the front wheels and slacken the rear wheel nuts, raise the rear of the vehicle, support it on stands and remove the road wheel. Withdraw the grease retainer cap from the centre of the hub and extract the split pin from the stub shaft. Unscrew the hub nut, remove the flat washer and withdraw the hub and brake drum assembly. Extract the hub oil seal, drive the inner bearing out and collect the spacer. Invert the hub and brake drum assembly and drive out the outer bearing. Inspect the bearings for signs of wear and damage.

▶ Pack the bearings with suitable grease and press the outer bearing into the hub with the side marked '*thrust*' facing outwards. Invert the hub, fit the spacer and press the inner bearing, with the side marked '*thrust*' outwards into the hub. Dip the new oil seal in oil and press it into the hub (sealing lip facing inwards). Refit all in reverse order.

Job card

Technician/learner name & date	Make and model	VIN no.		Reg. no.	Job/task no.

Customer's instructions/vehicle fault		Mileage		

Work carried out and recommendations (include PPE & special precautions taken)

Parts and labour	Price
Total	

Data and specifications used (include the actual figures)

Assessor report

	Assessment outcome	Passed (tick ✓)
1	The learner worked safely and minimised risks to themselves and others	
2	The learner correctly selected and used appropriate technical information	
3	The learner correctly selected and used appropriate tools and equipment	
4	The learner correctly carried out the task required using suitable methods and testing procedures	
5	The learner correctly recorded information and made suitable recommendations	

	Tick	Written feedback (with reference to assessment criteria) must be given when a learner is referred
Pass: I confirm that the learner's work was to an acceptable standard and met the assessment criteria of the unit		
Refer: The work carried out did not achieve the standards specified by the assessment criteria		

Assessor name (print)	Assessor PIN/ref.	Date

Section below only to be completed by the learner once the assessor decision has been made and feedback given			
I confirm that the work carried out was my own, and that I received feedback from the Assessor	Learner name (print)	Learner signature	Date

Worksheet 49: Check steering components

Procedure

▶ Steering wheel alignment – road test – spokes should be aligned when travelling straight ahead.

▶ Column security – rock up and down/side to side. Check adjustment is secure.

▶ Freeplay – movement of wheel without road wheels moving. Normally should be small (2.5 cm/1 in max.).

▶ Steering shaft and joints – side to side movement tests bearings. Rocking movement to test UJs.

▶ Rack/box security – check mounting bolts and rubber components.

▶ Rack gaiters/box seals – look for splits and oil leaks.

▶ Ball joints/track rod ends – as steering wheel is rocked, look for excessive 'lift' in the joints.

▶ Swivel joints/kingpins – jack and support vehicle. Rock wheel top and bottom. Look from the inside to check movement. Very little if any should be observed.

▶ Wheel bearings – as above.

Job card

Technician/learner name & date	Make and model	VIN no.	Reg. no.	Job/task no.

Customer's instructions/vehicle fault	Mileage	

Work carried out and recommendations (include PPE & special precautions taken)

Parts and labour	Price
Total	

Data and specifications used (include the actual figures)

Assessor report

Assessment outcome	Passed (tick ✓)
1	The learner worked safely and minimised risks to themselves and others	
2	The learner correctly selected and used appropriate technical information	
3	The learner correctly selected and used appropriate tools and equipment	
4	The learner correctly carried out the task required using suitable methods and testing procedures	
5	The learner correctly recorded information and made suitable recommendations	

	Tick	Written feedback (with reference to assessment criteria) must be given when a learner is referred
Pass: I confirm that the learner's work was to an acceptable standard and met the assessment criteria of the unit		
Refer: The work carried out did not achieve the standards specified by the assessment criteria		

Assessor name (print)	Assessor PIN/ref.	Date

Section below only to be completed by the learner once the assessor decision has been made and feedback given			
I confirm that the work carried out was my own, and that I received feedback from the Assessor	Learner name (print)	Learner signature	Date

Worksheet 50: Measure and adjust tracking (toe-in/out), using optical gauges

Procedure

▶ Carry out basic checks as described in the 'Check steering components' worksheet.

▶ Set the arms to the wheel size and then zero the tracking gauges by placing the tips of the arms together.

▶ Roll the vehicle back and forth to make sure the steering is not under stress.

▶ Position the gauges to the wheels and take a reading. Compare to specifications.

▶ If adjustment is required, check the position of the steering wheel. Adjustment is made by changing the overall length of the track rod.

▶ If the spokes are even, make equal adjustments to each end of the track rod.

▶ If the spokes are NOT even, turn the wheel until they are and then make adjustments to each end of the track rod such as to bring the wheels to the correct position.

▶ To adjust track rod length, undo the lock nuts and then turn the rod, which is threaded into the track rod end.

▶ Tighten the lock nuts and check the alignment again (from step 3).

▶ Carry out further adjustment if required – it is quite usual for accurate adjustment to need two or three changes.

▶ Secure all components and remove gauges.

Job card

Technician/learner name & date	Make and model	VIN no.		Reg. no.	Job/task no.

Customer's instructions/vehicle fault		Mileage	

Work carried out and recommendations (include PPE & special precautions taken)

	Price
Parts and labour	
Total	

Data and specifications used (include the actual figures)

Assessor report

	Assessment outcome	*Passed (tick ✓)*
1	The learner worked safely and minimised risks to themselves and others	
2	The learner correctly selected and used appropriate technical information	
3	The learner correctly selected and used appropriate tools and equipment	
4	The learner correctly carried out the task required using suitable methods and testing procedures	
5	The learner correctly recorded information and made suitable recommendations	

	Tick	Written feedback (with reference to assessment criteria) must be given when a learner is referred
Pass: I confirm that the learner's work was to an acceptable standard and met the assessment criteria of the unit		
Refer: The work carried out did not achieve the standards specified by the assessment criteria		

Assessor name (print)	Assessor PIN/ref.	Date

Section below only to be completed by the learner once the assessor decision has been made and feedback given			
I confirm that the work carried out was my own, and that I received feedback from the Assessor	Learner name (print)	Learner signature	Date

Worksheet 51: Service disc brakes and measure brake disc thickness and runout

Procedure

▶ Jack up and support the vehicle on stands or use a suitable hoist. Remove the appropriate wheels.

▶ Inspect the brake pads. Recommendations vary slightly but in most cases, the pads should be replaced if the lining is less than 1.5 mm. Methods of pad removal vary so check the manufacturer's data. However, most types are quite simple. The method described here relates to the type where part of the caliper is removed.

▶ Turn the steering to a lock position to allow easy access to the caliper and pads. Wash the caliper and pad assembly using a proprietary brake cleaner or suitable extractor.

▶ If necessary, remove some brake fluid from the reservoir. This is because when the piston is pushed back to allow new pads to be fitted, fluid can overflow. If a retaining bolt clip is fitted, it should be removed. Undo both caliper piston fixing bolts. Many types require an Allen key.

▶ Rock the assembly side to side. This moves the pads and pushes the piston in, just far enough to allow the caliper piston to be removed. Withdraw the pads. Use a small lever to help if a spring clip holds one of the pads into the piston. Keep the pads to show to the customer if necessary and then dispose of them in line with current regulations. Examine the disc for grooves and corrosion.

▶ Inspect the surface of the disc for signs of cracking and grooves. Small grooves are to be expected after a period of use. Grooves deeper than about 0.4 mm are usually considered excessive.

▶ Using a micrometer, measure the thickness of the disc at several different places around the disc, towards the centre and towards the outer edge.

▶ Compare the readings to the manufacturer's specifications. Some manufacturers stamp the minimum thickness just inside the centre of the disc.

▶ Mount a dial gauge (dial indicator) on a magnetic, or other appropriate type of stand, with the plunger running about 15 mm in from the outer edge of the disc.

▶ Zero the gauge and rotate the disc. Take note of changes in the dial gauge reading. Refinish a grooved disc if allowable. Consult manufacturer's recommendations. Refer to the manufacturer's specifications for maximum allowable run out. As a guide, 0.15 mm is usually considered the limit. Replace discs with excessive run out.

▶ Use a G/C clamp to push the caliper piston fully home. Fit the new pads in position together with anti-squeal shims if fitted. Some manufacturers recommend that copper grease be applied to the back and sides of each pad. Take care not to contaminate the lining. Repeat the process on the other side of the vehicle. Pads on both sides must always be replaced as a set.

▶ Refit the caliper and tighten all bolts to the recommended torque. Pump the brake pedal until it feels hard. This is to make sure the pads are moved fully into position. Double check correct fitment and then refit the road wheels and tighten wheel nuts to recommended torque. Lower the vehicle to the ground.

Job card

Technician/learner name & date	Make and model	VIN no.		Reg. no.	Job/task no.

Customer's instructions/vehicle fault	Mileage	

Work carried out and recommendations (include PPE & special precautions taken)

Parts and labour	Price
Total	

Data and specifications used (include the actual figures)

Assessor report

	Assessment outcome	Passed (tick ✓)
1	The learner worked safely and minimised risks to themselves and others	
2	The learner correctly selected and used appropriate technical information	
3	The learner correctly selected and used appropriate tools and equipment	
4	The learner correctly carried out the task required using suitable methods and testing procedures	
5	The learner correctly recorded information and made suitable recommendations	

	Tick	Written feedback (with reference to assessment criteria) must be given when a learner is referred
Pass: I confirm that the learner's work was to an acceptable standard and met the assessment criteria of the unit		
Refer: The work carried out did not achieve the standards specified by the assessment criteria		

Assessor name (print)	Assessor PIN/ref.	Date

Section below only to be completed by the learner once the assessor decision has been made and feedback given			
I confirm that the work carried out was my own, and that I received feedback from the Assessor	Learner name (print)	Learner signature	Date

Worksheet 52: Service drum brakes

Procedure

▶ Jack up and support the vehicle on stands or use a suitable hoist. Remove the appropriate wheels. Release the parking brake. Remove the cap that protects the hub nut. Remove the locking tab or pin if used. Undo the nut and remove the outer bearing. Remove the drum together with the inner bearing. *Or*, remove the drum fixing screw and remove the drum.

▶ Wash the backplate, shoes and drum assembly using a proprietary brake cleaner or suitable extractor. Inspect the brake shoes. Recommendations vary slightly, but in most cases, the shoes should be replaced if the lining is less than about 1.5 mm. Methods of shoe removal vary so check the manufacturer's data. Inspect brake drum for grooving. Refinish a grooved drum with a brake drum lathe if allowable. Consult manufacturer's recommendations. Remove the shoe hold-down fixings if fitted. These usually twist or pull free. Note the position of the shoe return springs and remove them with special brake spring tool if necessary. Remove the handbrake cable. On some vehicles, the shoes can be removed together with the handbrake cable, adjuster and return springs, which can then be taken off.

▶ Check the wheel cylinders for leaks by peeling back the dust seals. The cylinders should be overhauled or replaced if leaks are detected. Discard the old shoes in line with current regulations but keep them for the customer to examine if necessary. Clean off the backplate. Apply special grease to the shoe contact points. Note that ordinary grease will not stand the high temperatures. Fit the return springs and adjuster to the new shoes. Fit the shoes to the backplate, making sure they fit into the lower pivot and wheel cylinder slots. Use a shoe retractor to lever the shoes into place.

▶ Refit the handbrake cable and shoe hold-down clips. Make sure the shoes are centralized. Pre-adjust brake shoes and parking brake before installing brake drum/hub assembly and wheel bearing. Refit the drum, bearings and nut. Tighten to the correct torque. *Or*, refit the drum and fixing screw.

▶ Pump the brake pedal until it feels hard. This is to make sure the shoes are adjusted and moved fully into position. Check for correct fitment and that the drum spins freely, then refit the road wheels tightening to the correct torque. Lower the vehicle to the ground. Road test to ensure correct operation. Remember to check the handbrake operation and adjust the cable if necessary.

Job card

Technician/learner name & date	Make and model	VIN no.		Reg. no.	Job/task no.

Customer's instructions/vehicle fault		Mileage		

Work carried out and recommendations (include PPE & special precautions taken)

Parts and labour	Price
Total	

Data and specifications used (include the actual figures)

Assessor report

Assessment outcome		Passed (tick ✓)
1	The learner worked safely and minimised risks to themselves and others	
2	The learner correctly selected and used appropriate technical information	
3	The learner correctly selected and used appropriate tools and equipment	
4	The learner correctly carried out the task required using suitable methods and testing procedures	
5	The learner correctly recorded information and made suitable recommendations	

	Tick	Written feedback (with reference to assessment criteria) must be given when a learner is referred
Pass: I confirm that the learner's work was to an acceptable standard and met the assessment criteria of the unit		
Refer: The work carried out did not achieve the standards specified by the assessment criteria		

Assessor name (print)	Assessor PIN/ref.	Date

Section below only to be completed by the learner once the assessor decision has been made and feedback given			
I confirm that the work carried out was my own, and that I received feedback from the Assessor	Learner name (print)	Learner signature	Date

Worksheet 53: Remove, overhaul and refit brake caliper

Procedure

▶ Remove (note that fixing methods vary so refer to manufacturer's procedures):

- Raise and support the vehicle. Remove the wheel. Clean away any dust using a proprietary cleaning system. Undo the securing bolts and inspect them for wear or damage. Remove the brake pads. Clamp the flexible brake pipe using a proper pipe clamp. Undo the pipe from the caliper.

▶ Overhaul:

- To remove the piston from the caliper it is usually necessary to direct compressed air into the flexible pipe connection on the caliper. Use lots of paper or rags to catch lost fluid and protect the piston. It may be expelled with quite some force – take care. Remove the rubber dust seal if fitted.
- Depending on design, the piston seal will be part of the piston or part of the caliper cylinder. Note carefully how it is fitted and remove with a plastic or wooded tool.
- Inspect the piston and bore for signs of scratches, corrosion or excessive wear. If any serious damage is noted, the complete unit should be replaced. Light corrosion may be removed using a honing tool or very fine emery paper.
- Thoroughly clean all parts using brake fluid or a brake system cleaner. *Do not* use petroleum-based solvents. Cleanliness is very important. Lubricate the new piston seal with clean brake fluid and install it. Make sure it is fitted the correct way round. Refit the piston into the cylinder.

▶ Refit:

- Screw the flexible pipe into the caliper and refit the caliper mounting bolts. Make sure the pipe connection is secure. Refit the pads and secure in position as required. Remove the brake pipe clamp. The system will now require bleeding to remove air.
- This may be done using a pressure bleeder or a simple tube and bottle system! Where just one corner of the car has been disturbed, and a pipe clamp used, it is usually possible to just bleed that part. If the age of the brake fluid is unknown, it is advisable to flush the hydraulic system and bleed the complete system.

Note: brake fluid is hygroscopic, this means that it will absorb moisture from the atmosphere. Use only brake fluid from a new container or one that is known to have been recently opened.

- Connect a small pipe to the bleed nipple and place the other end into a clear bottle that is part full with clean brake fluid. Release the bleed nipple – about half a turn is usually enough. Get an assistant to pump the brake pedal slowly, whilst making sure the fluid reservoir remains topped off.
- Watch the bottle and when no more air is being expelled, get your assistant to hold the brake pedal down – and then tighten the bleed nipple.
- Make sure the reservoir is topped up to the correct level and check that the brake pedal feels hard when operated. Check for leaks, refit the wheel and lower the car to the ground.

Job card

Technician/learner name & date	Make and model	VIN no.		Reg. no.	Job/task no.

Customer's instructions/vehicle fault	Mileage	

Work carried out and recommendations (include PPE & special precautions taken)

Parts and labour	Price
Total	

Data and specifications used (include the actual figures)

Assessor report

	Assessment outcome	Passed (tick ✓)
1	The learner worked safely and minimised risks to themselves and others	
2	The learner correctly selected and used appropriate technical information	
3	The learner correctly selected and used appropriate tools and equipment	
4	The learner correctly carried out the task required using suitable methods and testing procedures	
5	The learner correctly recorded information and made suitable recommendations	

	Tick	Written feedback (with reference to assessment criteria) must be given when a learner is referred
Pass: I confirm that the learner's work was to an acceptable standard and met the assessment criteria of the unit		
Refer: The work carried out did not achieve the standards specified by the assessment criteria		

Assessor name (print)	Assessor PIN/ref.	Date

Section below only to be completed by the learner once the assessor decision has been made and feedback given			
I confirm that the work carried out was my own, and that I received feedback from the Assessor	Learner name (print)	Learner signature	Date

Worksheet 54: Remove and refit brake master cylinder

Procedure

▶ Removal:

- Remove fluid from reservoir and place rags to collect spillages. Remove any ancillary equipment so as to gain access to the master cylinder pipes and connections. Disconnect any warning light wires.

- Undo the brake pipe unions and make sure brake fluid does not contact paintwork. Most cylinders can now be removed by undoing the two main securing nuts or bolts (check specs). If necessary, remove the unit from the vehicle.

▶ Refit:

- If required by the manufacturer, bench bleed the replacement master cylinder. Use special pipes to connect the outlets back into the reservoir to do this. Refitting is now a reversal of the removing process. The system will now require bleeding to remove air. This may be done using a pressure bleeder or a simple tube and bottle system. If the age of the brake fluid is unknown, it is advisable to flush the hydraulic system and bleed the complete system.

Note: check the manufacturer's specs, because some systems should be bled in a particular sequence.

Note: brake fluid is hygroscopic, this means that it will absorb moisture from the atmosphere. Use only brake fluid from a new container or one that is known to have been recently opened. Do not store brake fluid for long periods of time.

- Connect a small pipe to the bleed nipple and place the other end into a clear bottle that is part full with clean brake fluid. Release the bleed nipple – about half a turn is usually enough. Get an assistant to pump the brake pedal slowly, whilst making sure the fluid reservoir remains topped off.

- Watch the bottle and when no more air is being expelled, get your assistant to hold the brake pedal down – and then tighten the bleed nipple.

- Make sure the reservoir is topped off to the correct level and check that the brake pedal feels hard when operated. Check for leaks and 'road' test to ensure correct operation.

Job card

Technician/learner name & date	Make and model	VIN no.		Reg. no.	Job/task no.

Customer's instructions/vehicle fault		Mileage			

Work carried out and recommendations (include PPE & special precautions taken)

Parts and labour	Price

Total	

Data and specifications used (include the actual figures)

Assessor report

	Assessment outcome	Passed (tick ✓)
1	The learner worked safely and minimised risks to themselves and others	
2	The learner correctly selected and used appropriate technical information	
3	The learner correctly selected and used appropriate tools and equipment	
4	The learner correctly carried out the task required using suitable methods and testing procedures	
5	The learner correctly recorded information and made suitable recommendations	

	Tick	Written feedback (with reference to assessment criteria) must be given when a learner is referred
Pass: I confirm that the learner's work was to an acceptable standard and met the assessment criteria of the unit		
Refer: The work carried out did not achieve the standards specified by the assessment criteria		

Assessor name (print)	Assessor PIN/ref.	Date

Section below only to be completed by the learner once the assessor decision has been made and feedback given			
I confirm that the work carried out was my own, and that I received feedback from the Assessor	Learner name (print)	Learner signature	Date

Worksheet 55: Check vacuum servo unit operation

Procedure

▶ Ensure that the engine is in good running order because the vacuum servo is powered by a connection to the intake manifold. Check for air leaks as foot pressure is applied.

▶ Check servo operation by applying pressure to the pedal and starting the engine. Your foot should move further down as the engine starts and servo assistance is applied. It may not be necessary to check further if this test result is satisfactory. However, a thorough test is always advisable.

▶ Shut off the engine and wait five minutes. Apply the brakes and check that servo assistance is available for at least one application. This indicates a good air seal, if vacuum is retained.

▶ Connect a vacuum gauge to the inlet manifold and note the reading with the engine running at idle. A reading of about 0.5 bar is typical.

▶ Connect a vacuum gauge to the servo pipe, after the check valve, and note the reading with the engine running at idle. The reading should be the same as before. If not, check the pipe for kinks, blockages.

▶ Make sure the check valve blocks when you blow from the manifold end. Replace if in any doubt.

▶ Reconnect all parts and check for leaks.

Job card

Technician/learner name & date	Make and model	VIN no.		Reg. no.	Job/task no.

Customer's instructions/vehicle fault		Mileage		

Work carried out and recommendations (include PPE & special precautions taken)

Parts and labour	Price
Total	

Data and specifications used (include the actual figures)

Assessor report

Assessment outcome		Passed (tick ✓)
1	The learner worked safely and minimised risks to themselves and others	
2	The learner correctly selected and used appropriate technical information	
3	The learner correctly selected and used appropriate tools and equipment	
4	The learner correctly carried out the task required using suitable methods and testing procedures	
5	The learner correctly recorded information and made suitable recommendations	

	Tick	Written feedback (with reference to assessment criteria) must be given when a learner is referred
Pass: I confirm that the learner's work was to an acceptable standard and met the assessment criteria of the unit		
Refer: The work carried out did not achieve the standards specified by the assessment criteria		

Assessor name (print)	Assessor PIN/ref.	Date

Section below only to be completed by the learner once the assessor decision has been made and feedback given			
I confirm that the work carried out was my own, and that I received feedback from the Assessor	Learner name (print)	Learner signature	Date

Worksheet 56: Inspect, test and replace brake warning lights

Procedure

▶ Check stop light operation and adjust switch position. Remove the two wires from the stoplight switch and bridge them together with a fused jumper wire. The ignition may need to be switched on. The stoplights should light. If not, trace the circuit for a break starting with the fuse.

▶ If the lights work when the switch is bridged, the switch needs replacing or adjusting. Most switches are positioned above the brake pedal and have a screwed body with adjusting nuts. These switches make contact as the plunger springs out. Adjust the switch position so that the lug on the brake pedal allows the plunger to move as soon as the pedal is pushed down. Check that the switch is not too sensitive, such that the lights flash on due to vibration for example. Secure all wires and adjusting nuts – check operation again.

▶ Check parking brake indicator light/audible warning – this is a generic test routine; refer to manufacturer's procedures and circuits for specific details. Apply the handbrake. Switch the ignition on but do not start the engine. Check that the parking brake warning light is illuminated on the dashboard. If the warning light does not illuminate, check switch adjustment, if OK, simply bridging the switch contact will confirm a sound circuit. Release the handbrake and check the warning light extinguishes. If not, check switch adjustment.

▶ For vehicles fitted with an audible warning device, start the engine and drive vehicle from its position of rest with the handbrake partially applied. The audible warning device should be heard. Check brake warning light – this is a generic test routine; refer to manufacturer's procedures and circuits for specific details. Release parking brake and switch ignition.

▶ Check that the general brake warning light is not illuminated on the dashboard. If illuminated, check front brake pads for wear, and the master cylinder fluid level. Correct as appropriate.

Job card

Technician/learner name & date	Make and model	VIN no.		Reg. no.	Job/task no.

Customer's instructions/vehicle fault		Mileage			

Work carried out and recommendations (include PPE & special precautions taken)

Parts and labour		Price

Total		

Data and specifications used (include the actual figures)

Assessor report

	Assessment outcome	Passed (tick ✓)
1	The learner worked safely and minimised risks to themselves and others	
2	The learner correctly selected and used appropriate technical information	
3	The learner correctly selected and used appropriate tools and equipment	
4	The learner correctly carried out the task required using suitable methods and testing procedures	
5	The learner correctly recorded information and made suitable recommendations	

	Tick	Written feedback (with reference to assessment criteria) must be given when a learner is referred
Pass: I confirm that the learner's work was to an acceptable standard and met the assessment criteria of the unit		
Refer: The work carried out did not achieve the standards specified by the assessment criteria		

Assessor name (print)	Assessor PIN/ref.	Date

Section below only to be completed by the learner once the assessor decision has been made and feedback given			
I confirm that the work carried out was my own, and that I received feedback from the Assessor	Learner name (print)	Learner signature	Date

Worksheet 57: Measure tyre tread and report on condition

Procedure

▶ Check: NSF, OSF, NSR, OSR.

▶ Check and report minimum tread depth.

▶ Check and report condition.

▶ Check suitability for vehicle.

▶ Manufacturer's wheel nut torque.

▶ Set wheel nut torque.

▶ Manufacturer's pressure.

▶ Check/adjust pressure.

▶ Report to customer.

Job card

Technician/learner name & date	Make and model	VIN no.	Reg. no.	Job/task no.

Customer's instructions/vehicle fault	Mileage	

Work carried out and recommendations (include PPE & special precautions taken)

Parts and labour	Price
Total	

Data and specifications used (include the actual figures)

Assessor report

	Assessment outcome	Passed (tick ✓)
1	The learner worked safely and minimised risks to themselves and others	
2	The learner correctly selected and used appropriate technical information	
3	The learner correctly selected and used appropriate tools and equipment	
4	The learner correctly carried out the task required using suitable methods and testing procedures	
5	The learner correctly recorded information and made suitable recommendations	

	Tick	Written feedback (with reference to assessment criteria) must be given when a learner is referred
Pass: I confirm that the learner's work was to an acceptable standard and met the assessment criteria of the unit		
Refer: The work carried out did not achieve the standards specified by the assessment criteria		

Assessor name (print)	Assessor PIN/ref.	Date

Section below only to be completed by the learner once the assessor decision has been made and feedback given			
I confirm that the work carried out was my own, and that I received feedback from the Assessor	Learner name (print)	Learner signature	Date

Transmission

Worksheet 58: Remove and refit gear change mechanism

Procedure

▶ Remove:

- Disconnect the earth/ground cable from the battery.

- On some cars, it will be necessary to remove the passenger's or driver's seat. Disconnect any electrical wires.

- Remove the knob at the top of the gear lever. Most will unscrew but some are held with a small screw.

- Remove console covers, gaiters and panels as necessary to gain access.

- Disconnect the shift rod levers or unscrew the ball joint cover as appropriate.

- Disconnect the electrical wiring for the overdrive switch (if fitted).

- Remove the gear lever.

▶ Refit:

- Place the gear lever in position and install the ball joint cover or shift rod levers.

- Adjust the linkage if necessary.

- Connect the overdrive electrical switch wiring (if disconnected).

- Reinstall the covers, panels and other components as appropriate.

Job card

Technician/learner name & date	Make and model	VIN no.		Reg. no.	Job/task no.

Customer's instructions/vehicle fault	Mileage	

Work carried out and recommendations (include PPE & special precautions taken)

Parts and labour	Price
Total	

Data and specifications used (include the actual figures)

Assessor report

	Assessment outcome	Passed (tick ✓)
1	The learner worked safely and minimised risks to themselves and others	
2	The learner correctly selected and used appropriate technical information	
3	The learner correctly selected and used appropriate tools and equipment	
4	The learner correctly carried out the task required using suitable methods and testing procedures	
5	The learner correctly recorded information and made suitable recommendations	

	Tick	Written feedback (with reference to assessment criteria) must be given when a learner is referred
Pass: I confirm that the learner's work was to an acceptable standard and met the assessment criteria of the unit		
Refer: The work carried out did not achieve the standards specified by the assessment criteria		

Assessor name (print)	Assessor PIN/ref.	Date

Section below only to be completed by the learner once the assessor decision has been made and feedback given			
I confirm that the work carried out was my own, and that I received feedback from the Assessor	Learner name (print)	Learner signature	Date

Worksheet 59: Remove and refit transmission gearbox (transaxle type)

Procedure

▶ Remove:

- Support the vehicle on a suitable hoist. Fit car protection kit as required and disconnect the battery. Remember to fit a memory keeper if necessary. Drain the gearbox oil.

- Remove any ancillary components as necessary that allow easier access – the exhaust for example.

- Mark the gear change linkage and then remove parts as required. Remove the minimum number of parts or remove the linkage as a complete unit where possible. This makes refitting easier.

- On some vehicles, it is necessary to remove the suspension on one side to allow access to the gearbox, and for removal of the driveshafts. Remove the driveshafts (a separate worksheet is available).

- Remove the speedometer cable or speed sensor. Remove the reverse (backup) light switch wires. Tie these components out of the way if necessary. Remove the starter motor if necessary.

- Fit engine support bar as required; remove and inspect mountings, and cross members. Support the gearbox on a transmission jack if necessary and remove the gearbox or bell housing bolts.

- Move the gearbox straight out of the clutch assembly, away from the vehicle and place on a suitable bench.

▶ Refit:

- As usual – refitting is a reversal of the removal process! However, it is normal to remove and check the clutch assembly. When this is refitted, make sure it is aligned correctly because this makes refitting the gearbox much easier.

- Remember to refill with the correct lubricant and that all fixings are tightened correctly.

- A road test is recommended to ensure correct operation when the job is completed.

Job card

Technician/learner name & date	Make and model	VIN no.		Reg. no.	Job/task no.

Customer's instructions/vehicle fault	Mileage	

Work carried out and recommendations (include PPE & special precautions taken)

Parts and labour	Price
Total	

Data and specifications used (include the actual figures)

Assessor report

	Assessment outcome	Passed (tick ✓)
1	The learner worked safely and minimised risks to themselves and others	
2	The learner correctly selected and used appropriate technical information	
3	The learner correctly selected and used appropriate tools and equipment	
4	The learner correctly carried out the task required using suitable methods and testing procedures	
5	The learner correctly recorded information and made suitable recommendations	

	Tick	Written feedback (with reference to assessment criteria) must be given when a learner is referred
Pass: I confirm that the learner's work was to an acceptable standard and met the assessment criteria of the unit		
Refer: The work carried out did not achieve the standards specified by the assessment criteria		

Assessor name (print)	Assessor PIN/ref.	Date

Section below only to be completed by the learner once the assessor decision has been made and feedback given			
I confirm that the work carried out was my own, and that I received feedback from the Assessor	Learner name (print)	Learner signature	Date

Worksheet 60: Replace clutch assembly

Procedure

▶ Obtain and follow manufacturer's procedures. Disconnect battery (ground first), fitting a memory keeper if necessary. Raise vehicle. Remove starter and driveline components.

▶ Remove clutch cable/slave cylinder and transmission – following procedures.

▶ Remove clutch fork and release bearing assembly. Remove any dust using recommended health procedures.

▶ Mark the clutch cover if necessary and remove the ring of bolts.

▶ Remove clutch cover and plate.

▶ Inspect, repair, replace as necessary (it is usual to replace as a set). Inspect engine block, clutch (bell) housing and transmission/transaxle case mating surfaces. Inspect, remove or replace crankshaft pilot bearing or bushing as applicable.

▶ Check condition of flywheel using dial gauge (see next worksheet). Check ring gear for wear and cracks (measure ring gear run out). Measure crankshaft end play.

▶ Using alignment tool, replace new disc and cover.

▶ Refit all in reverse procedure.

▶ Bleed clutch hydraulic system. Test operation. Road test and report findings.

Job card

Technician/learner name & date	Make and model	VIN no.	Reg. no.	Job/task no.

Customer's instructions/vehicle fault	Mileage	

Work carried out and recommendations (include PPE & special precautions taken)

Parts and labour	Price
Total	

Data and specifications used (include the actual figures)

Assessor report

Assessment outcome		Passed (tick ✓)
1	The learner worked safely and minimised risks to themselves and others	
2	The learner correctly selected and used appropriate technical information	
3	The learner correctly selected and used appropriate tools and equipment	
4	The learner correctly carried out the task required using suitable methods and testing procedures	
5	The learner correctly recorded information and made suitable recommendations	

	Tick	Written feedback (with reference to assessment criteria) must be given when a learner is referred
Pass: I confirm that the learner's work was to an acceptable standard and met the assessment criteria of the unit		
Refer: The work carried out did not achieve the standards specified by the assessment criteria		

Assessor name (print)	Assessor PIN/ref.	Date

Section below only to be completed by the learner once the assessor decision has been made and feedback given			
I confirm that the work carried out was my own, and that I received feedback from the Assessor	Learner name (print)	Learner signature	Date

Worksheet 61: Remove and refit wheel bearings

Procedure

▶ Front hub assembly:

- Apply the handbrake and slacken the front road wheel nuts, raise the front of the vehicle, support it on stands and remove the road wheel. Remove the drive shaft nut split pin, use an assistant to apply firm pressure to the brake pedal and, while the brake is applied, unscrew the drive shaft nut. Remove the brake caliper and support it to prevent straining the hose. Remove the disc.

- Using a ball joint breaker tool, disconnect the ball joint from the steering lever. Unscrew the nuts and remove the bolts to release the strut from the hub assembly.

- Unscrew the nut and remove the clamp bolt securing the lower ball joint to the hub assembly and, using a suitable lever placed between the lower arm and the anti-roll bar, lever downwards to release the ball joint from the hub. Remove the hub assembly from the drive shaft.

- Extract the inner oil seal and spacer and the outer oil seal. Drive out one of the bearings, invert the hub and drive out the remaining bearing. Inspect the bearings for signs of wear and damage, renew as necessary. Pack the bearings with suitable grease and press them into the hub. Fit the oil seal spacer against the inner bearing. Fit the oil seals.

- Locate the hub on the drive shaft, fit the flat washer and drive shaft nut and tighten the nut finger tight. Fit the hub assembly to the lower ball joint, fit the clamp bolt and tighten the nut. Fit the hub to the strut, fit the bolts and tighten the nuts to the correct torque.

- Connect the ball joint to the steering lever and fit and tighten the nut. Fit the disc to the drive flange and tighten the securing screws. Fit the brake caliper. Use an assistant to apply firm pressure to the brake pedal and, while the brake is applied, tighten the drive shaft nut to the correct torque. Lock the nut with a new split pin. Fit the road wheel and nuts.

▶ Rear hub assembly:

- Chock the front wheels and slacken the rear wheel nuts, raise the rear of the vehicle, support it on stands and remove the road wheel. Withdraw the grease retainer cap from the centre of the hub and extract the split pin from the stub shaft.

- Unscrew the hub nut, remove the flat washer and withdraw the hub and brake drum assembly. Extract the hub oil seal, drive the inner bearing out and collect the spacer. Invert the hub and brake drum assembly and drive out the outer bearing. Inspect the bearings for signs of wear and damage, renew as necessary.

- Pack the bearings with suitable grease and press the outer bearing into the hub with the side marked '*thrust*' facing outwards. Invert the hub, fit the spacer and press the inner bearing, with the side marked '*thrust*' outwards into the hub. Dip the new oil seal in oil and press it into the hub (sealing lip facing inwards).

- Fit the hub and brake drum assembly to the stub shaft, fit the flat washer and fit and tighten the hub nut to the correct torque. Lock the nut with a new split pin. Measure drive axle flange run out and shaft endplay. Fit the grease retainer cap, then fit the road wheel and nuts.

Job card

Technician/learner name & date	Make and model	VIN no.		Reg. no.	Job/task no.

Customer's instructions/vehicle fault	Mileage	

Work carried out and recommendations (include PPE & special precautions taken)

	Price
Parts and labour	
Total	

Data and specifications used (include the actual figures)

Assessor report

	Assessment outcome	Passed (tick ✓)
1	The learner worked safely and minimised risks to themselves and others	
2	The learner correctly selected and used appropriate technical information	
3	The learner correctly selected and used appropriate tools and equipment	
4	The learner correctly carried out the task required using suitable methods and testing procedures	
5	The learner correctly recorded information and made suitable recommendations	

	Tick	Written feedback (with reference to assessment criteria) must be given when a learner is referred
Pass: I confirm that the learner's work was to an acceptable standard and met the assessment criteria of the unit		
Refer: The work carried out did not achieve the standards specified by the assessment criteria		

Assessor name (print)	Assessor PIN/ref.	Date

Section below only to be completed by the learner once the assessor decision has been made and feedback given			
I confirm that the work carried out was my own, and that I received feedback from the Assessor	Learner name (print)	Learner signature	Date

Worksheet 62: Remove and refit driveshaft

Procedure

▶ Apply the handbrake and slacken the road wheel nuts.

▶ Raise the front of the vehicle, support it on stands and remove the road wheel.

▶ Remove the driveshaft nut split pin or lock tab.

▶ Use an assistant to apply the footbrake and then remove the driveshaft nut and washer.

▶ Split the steering track rod end from the steering arm and remove it.

▶ Remove the bolts securing the hub to the suspension strut.

▶ Pivot the hub outwards to the limit of its movement but take care not to strain the brake hose.

▶ Manoeuvre the drive shaft from the hub.

▶ Carefully lever between the driveshaft inner joint and the differential housing to release the spring ring. Withdraw the driveshaft.

▶ To refit, slide the shaft into the differential housing until the spring ring engages.

▶ Manoeuvre the outer end of the drive shaft into the hub and fit the nut and washer. A new nut may be required by some manufacturers.

▶ Refit the suspension strut and the steering joint.

▶ Use an assistant to apply the footbrake and then tighten the driveshaft nut to the specified torque. Fit a new split pin or knock in the tab as required.

▶ Refit the road wheel and lower the vehicle. Torque the wheel nuts and road test.

Job card

Technician/learner name & date	Make and model	VIN no.		Reg. no.	Job/task no.

Customer's instructions/vehicle fault		Mileage	

Work carried out and recommendations (include PPE & special precautions taken)

Parts and labour	Price
Total	

Data and specifications used (include the actual figures)

Assessor report

	Assessment outcome	Passed (tick ✓)
1	The learner worked safely and minimised risks to themselves and others	
2	The learner correctly selected and used appropriate technical information	
3	The learner correctly selected and used appropriate tools and equipment	
4	The learner correctly carried out the task required using suitable methods and testing procedures	
5	The learner correctly recorded information and made suitable recommendations	

	Tick	Written feedback (with reference to assessment criteria) must be given when a learner is referred
Pass: I confirm that the learner's work was to an acceptable standard and met the assessment criteria of the unit		
Refer: The work carried out did not achieve the standards specified by the assessment criteria		

Assessor name (print)	Assessor PIN/ref.	Date

Section below only to be completed by the learner once the assessor decision has been made and feedback given			
I confirm that the work carried out was my own, and that I received feedback from the Assessor	Learner name (print)	Learner signature	Date

Worksheet 63: Remove and refit final drive and differential (FWD and RWD)

Procedure

▶ Remove FWD:

- Drain the oil from the gearbox and refit the drain plug. Remove the gearbox from the vehicle. Position the gearbox on its bell housing face and remove the gear case, selector shafts and forks, the mainshaft and countershaft. Lift out the final drive gear and differential assembly.

- Remove the roller bearing from the bell housing. Remove the carrier bearings. Remove the differential oil seals. Remove the bolts securing the final drive gear to the differential housing and withdraw the final drive gear.

- Remove the roll pin securing the differential pinion shaft and remove the pinion shaft. Remove the planet gears, thrust washers and the sun gears. Clean all components and examine for wear and damage.

▶ Refit FWD:

- Fit the pinion roller bearing to the bell housing. Lubricate and fit the sun gears and planet gears and the differential pinion shaft. Ensure that the roll pinhole is aligned with the differential housing and fit a new roll pin. Select a thrust washer, which will provide the correct backlash.

- Thrust washer dimensions must be equal in both gears. Backlash may be checked using the vehicle drive shaft inner couplings to centralize the sun gears.

- Ensure that the mating faces of the final drive gear and the differential housing are clean and free of burrs. Fit the final drive gear and secure the bolts to the correct torque. Fit the carrier bearings. Fit the final drive and differential assembly to the bell housing. Fit the gear case and secure it to the bell housing.

- Ensure that the final drive assembly is fully in position in the bell housing. Check the clearance between the gear case carrier bearing and the gear case bearing recess.

- If there is no clearance or excessive clearance, slacken the gear case bolts and remove the circlip type shim through the oil seal location. Substitute an alternative shim as required, tighten the gear case bolts and recheck final drive end float.

- Fit the differential oil seals using special tools as required. Lubricate the seal lips. Fit the gearbox components. Apply sealant to the gear case face and fit the gear case. Fit the gearbox to the vehicle and fill with the correct grade and quantity of oil.

▶ Remove RWD:

- Jack up and support on stands. Remove the wheels. Undo and remove brake drums. Inspect wheel studs and replace if threads are worn or damaged.

- Unscrew the bolts holding the bearing clamp and pull out the halfshafts. A slide hammer may be required. Remove the propshaft.

- Drain oil from the unit if possible – or use a tray to catch the oil as the whole unit is removed. Undo the ring of bolts around the final drive housing. Remove the final drive assembly – with assistance if necessary. Clean off any old gaskets. Inspect and replace seals. Renew gaskets and use sealant as required.

▶ Refit RWD:

- This is a reversal of the removal process. Top up with the correct new oil. Measure flange run-out and shaft endplay.

Job card

Technician/learner name & date	Make and model	VIN no.		Reg. no.	Job/task no.
Customer's instructions/vehicle fault		**Mileage**			

Work carried out and recommendations (include PPE & special precautions taken)	

Parts and labour	**Price**
Total	

Data and specifications used (include the actual figures)

Assessor report

Assessment outcome		Passed (tick ✓)
1	The learner worked safely and minimised risks to themselves and others	
2	The learner correctly selected and used appropriate technical information	
3	The learner correctly selected and used appropriate tools and equipment	
4	The learner correctly carried out the task required using suitable methods and testing procedures	
5	The learner correctly recorded information and made suitable recommendations	

	Tick	Written feedback (with reference to assessment criteria) must be given when a learner is referred
Pass: I confirm that the learner's work was to an acceptable standard and met the assessment criteria of the unit		
Refer: The work carried out did not achieve the standards specified by the assessment criteria		

Assessor name (print)	Assessor PIN/ref.	Date

Section below only to be completed by the learner once the assessor decision has been made and feedback given			
I confirm that the work carried out was my own, and that I received feedback from the Assessor	Learner name (print)	Learner signature	Date

Worksheet 64: Inspect and measure backlash, final drive tooth wear and pinion torque

Procedure

▶ Remove final drive and differential assembly from the vehicle. Note that the figures listed here are typical but always refer to data specific to the vehicle.

▶ Backlash describes the movement of the crown wheel before it contacts and moves the pinion. It is adjusted by setting the position of the two main bearings.

▶ Tighten bearing cap bolts and slacken off again. Then tighten the cap bolts finger tight.

▶ Screw the two adjusting nuts, with a special tool if necessary, lightly against the bearings.

▶ Set a dial gauge on a magnetic stand and against one tooth of the crown wheel.

▶ Tighten the adjusting nut on the crown wheel side until a backlash of 0.01 mm is obtained.

▶ Next, preload the bearing on the differential side.

▶ Measure the backlash at four opposing points and adjust the nut until a reading of 0.1 to 0.2 mm is obtained.

▶ Spin the pinion gear several times, recheck and tighten the bearing caps to the prescribed torque.

▶ Coat the crown wheel teeth with touch-up paint or 'engineer's blue'. Spin the drive flange several times while braking the crown wheel with a hardwood wedge. Check the wear pattern and adjust the backlash as required within the specified limits as necessary.

▶ Fit lock tabs to the main adjusting nuts.

▶ Finally, measure the drive pinion turning torque. If this is incorrect, a new collapsible spacer must be fitted and the pinion nut torque set. Alternatively, shims are used to set the pinion.

▶ Refit to the vehicle and top up with oil. Use new gaskets as required.

Job card

Technician/learner name & date	Make and model	VIN no.	Reg. no.	Job/task no.

Customer's instructions/vehicle fault	Mileage	

Work carried out and recommendations (include PPE & special precautions taken)

Parts and labour	Price
Total	

Data and specifications used (include the actual figures)

Assessor report

	Assessment outcome		Passed (tick ✓)
1	The learner worked safely and minimised risks to themselves and others		
2	The learner correctly selected and used appropriate technical information		
3	The learner correctly selected and used appropriate tools and equipment		
4	The learner correctly carried out the task required using suitable methods and testing procedures		
5	The learner correctly recorded information and made suitable recommendations		

	Tick	Written feedback (with reference to assessment criteria) must be given when a learner is referred
Pass: I confirm that the learner's work was to an acceptable standard and met the assessment criteria of the unit		
Refer: The work carried out did not achieve the standards specified by the assessment criteria		

Assessor name (print)	Assessor PIN/ref.	Date

Section below only to be completed by the learner once the assessor decision has been made and feedback given			
I confirm that the work carried out was my own, and that I received feedback from the Assessor	Learner name (print)	Learner signature	Date

Worksheet 65: Remove and replace propshaft and replace universal joint (UJ) and centre bearing

Procedure

▶ Removal:

- Jack up vehicle and support on axle stands. If necessary disconnect exhaust to improve access and removal of propshaft. Mark the rear UJ flange and final drive flange in relation to each other. Remove bolts securing the propshaft to the final drive. Remove bolts securing the centre bearing to its support bracket on the underbody, noting the location and number of slotted shims between the bearing and bracket.

- Disconnect the propshaft front end from the gearbox by pulling the shaft rearwards. To prevent loss of transmission oil, a suitable plug should be inserted into the oil seal.

▶ Replacement of universal joint:

Note: not all universal joints are replaceable. If not, a replacement propshaft will need to be obtained. Clean away all traces of dirt from the four UJ circlips and remove. If they are difficult to remove, tap the bearing surface resting on the spider with a mallet.

- Remove the bearing cups and needle rollers by tapping the yoke at each bearing with a mallet. With the bearing cups removed, extract the spider from the yokes. Thoroughly clean the yokes, journals and circlip grooves. Fit the new spider on the yoke. Refit the bearing cups on the spider and press the bearings home using a press or vice. Replace the circlips.

▶ Replacement of centre bearing

- Mark the front and rear sections of the propshaft in relation to each other, and also mark the position of the U-shaped washer located beneath the head of the bolt in the centre UJ.

- Loosen the bolt in the central UJ so that the U-shaped washer can be removed. With the U-shaped washer removed, slide the rear section of the propshaft from the front. Remove the centre bearing using a suitable puller. Refitting is a reversal of removal. Ensure the U-shaped washer is fitted in the previously marked position.

▶ Refitting the propshaft

- Refitting is a reversal of removal. Ensure that previously noted slotted shims are returned to their original location. Using a straight edge of suitable length, check that the two sections of the propshaft are aligned. Check manufacturer's data for maximum joint angles. Adjustment will be by adding or removing shims from the centre bearing. Check transmission fluid level.

Job card

Technician/learner name & date	Make and model	VIN no.	Reg. no.	Job/task no.

Customer's instructions/vehicle fault		Mileage	

Work carried out and recommendations (include PPE & special precautions taken)

	Price
Parts and labour	
Total	

Data and specifications used (include the actual figures)

Assessor report

	Assessment outcome	Passed (tick ✓)
1	The learner worked safely and minimised risks to themselves and others	
2	The learner correctly selected and used appropriate technical information	
3	The learner correctly selected and used appropriate tools and equipment	
4	The learner correctly carried out the task required using suitable methods and testing procedures	
5	The learner correctly recorded information and made suitable recommendations	

	Tick	Written feedback (with reference to assessment criteria) must be given when a learner is referred
Pass: I confirm that the learner's work was to an acceptable standard and met the assessment criteria of the unit		
Refer: The work carried out did not achieve the standards specified by the assessment criteria		

Assessor name (print)	Assessor PIN/ref.	Date

Section below only to be completed by the learner once the assessor decision has been made and feedback given

I confirm that the work carried out was my own, and that I received feedback from the Assessor	Learner name (print)	Learner signature	Date

Electrical

Worksheet 66: Remove and refit electrical components

Procedure

▶ Check data for location of fuse box.

▶ Switch on the circuit under test and the ignition if necessary.

▶ Check for voltage at both sides of the fuse, fusible link or circuit breaker with a meter or a test lamp. Ceramic and glass fuses have metal contacts; blade fuses have small test points at each side on top of the fuse.

▶ Remove the fuse using a clip puller if necessary. Replace with one of the recommended value.

▶ Check bulb replacement methods. Most rear lights are accessible from behind the light cluster inside the vehicle. Headlight bulbs usually have a small cover that should be removed. Side and indicator light units are often removed as a unit after a spring clip is released.

▶ Check bulbs by eye but also with an ohmmeter – a few ohms will indicate that the bulb is in good order. Do not touch the glass of a headlight bulb. This can create a localized hot spot due to grease contamination and cause the bulb to blow. Clean off carefully if touched accidentally.

▶ Replace faulty bulbs with ones of the correct rating (voltage and wattage) as well as the correct fitting. In some cases, the bulbs are coloured.

▶ Check all lights for correct operation.

Job card

Technician/learner name & date	Make and model	VIN no.		Reg. no.	Job/task no.
Customer's instructions/vehicle fault		Mileage			

Work carried out and recommendations (include PPE & special precautions taken)

Parts and labour	Price
Total	

Data and specifications used (include the actual figures)

Assessor report

Assessment outcome		Passed (tick ✓)
1	The learner worked safely and minimised risks to themselves and others	
2	The learner correctly selected and used appropriate technical information	
3	The learner correctly selected and used appropriate tools and equipment	
4	The learner correctly carried out the task required using suitable methods and testing procedures	
5	The learner correctly recorded information and made suitable recommendations	

	Tick	Written feedback (with reference to assessment criteria) must be given when a learner is referred
Pass: I confirm that the learner's work was to an acceptable standard and met the assessment criteria of the unit		
Refer: The work carried out did not achieve the standards specified by the assessment criteria		

Assessor name (print)	Assessor PIN/ref.	Date

Section below only to be completed by the learner once the assessor decision has been made and feedback given			
I confirm that the work carried out was my own, and that I received feedback from the Assessor	Learner name (print)	Learner signature	Date

Worksheet 67: Remove and refit electrical components

Procedure

▶ This worksheet is generic and can be applied to many systems. However, refer to manufacturer's procedures for specific information.

▶ Fit a memory keeper to prevent changing stored settings in electronic control units and in car entertainment systems.

▶ Remove the battery earth/ground lead.

▶ If removing the battery, disconnect the supply connection, remove the casing clamp and remove the battery. Take care to keep it level so that no electrolyte (sulphuric acid) is spilt.

▶ For other components, disconnect their supply wires. Making a note or suitable sketch where necessary of the connections will save time when refitting.

▶ For some components, it may be necessary to remove other parts to allow easy access. For example, an exhaust shield may need to be removed before the alternator can be disconnected.

▶ Disconnect linkages and/or peripherals.

▶ Undo all mountings and remove the unit from the vehicle.

▶ Refitting is a reversal of the removal process.

▶ The last job is always to reconnect the battery earth/ground lead.

Job card

Technician/learner name & date	Make and model	VIN no.		Reg. no.	Job/task no.

Customer's instructions/vehicle fault		Mileage		

Work carried out and recommendations (include PPE & special precautions taken)

Parts and labour	Price
Total	

Data and specifications used (include the actual figures)

Assessor report

Assessment outcome		Passed (tick ✓)
1	The learner worked safely and minimised risks to themselves and others	
2	The learner correctly selected and used appropriate technical information	
3	The learner correctly selected and used appropriate tools and equipment	
4	The learner correctly carried out the task required using suitable methods and testing procedures	
5	The learner correctly recorded information and made suitable recommendations	

	Tick	Written feedback (with reference to assessment criteria) must be given when a learner is referred
Pass: I confirm that the learner's work was to an acceptable standard and met the assessment criteria of the unit		
Refer: The work carried out did not achieve the standards specified by the assessment criteria		

Assessor name (print)	Assessor PIN/ref.	Date

Section below only to be completed by the learner once the assessor decision has been made and feedback given			
I confirm that the work carried out was my own, and that I received feedback from the Assessor	Learner name (print)	Learner signature	Date

Worksheet 68: Remove and replace headlight unit

Procedure

▶ Fit memory keeper and disconnect the battery earth/ground lead.

▶ Remove covers as required and disconnect wires to lights.

▶ Remove grill and/or trim as necessary for access to light unit fixings.

▶ Undo bolts and/or clips and remove light unit from the car.

▶ Refitting is a reverse of the removal process.

▶ Note that some manufacturers require special contact grease to be applied to the terminals. This makes for a good electrical contact and keeps water out.

▶ Check and adjust alignment.

▶ Check all other lights.

Job card

Technician/learner name & date	Make and model	VIN no.		Reg. no.	Job/task no.
Customer's instructions/vehicle fault		**Mileage**			
Work carried out and recommendations (include PPE & special precautions taken)					
Parts and labour					**Price**
Total					
Data and specifications used (include the actual figures)					

Assessor report

	Assessment outcome	Passed (tick ✓)
1	The learner worked safely and minimised risks to themselves and others	
2	The learner correctly selected and used appropriate technical information	
3	The learner correctly selected and used appropriate tools and equipment	
4	The learner correctly carried out the task required using suitable methods and testing procedures	
5	The learner correctly recorded information and made suitable recommendations	

	Tick	Written feedback (with reference to assessment criteria) must be given when a learner is referred
Pass: I confirm that the learner's work was to an acceptable standard and met the assessment criteria of the unit		
Refer: The work carried out did not achieve the standards specified by the assessment criteria		

Assessor name (print)	Assessor PIN/ref.	Date

Section below only to be completed by the learner once the assessor decision has been made and feedback given			
I confirm that the work carried out was my own, and that I received feedback from the Assessor	**Learner name (print)**	**Learner signature**	**Date**

Worksheet 69: Remove and refit flasher unit

Procedure

▶ Note that if the left or right indicators, or the hazards work, the unit is probably functioning correctly. Remember that a flasher unit is designed to flash at a different rate when a bulb is blown.

▶ Flasher units are usually located either as part of the fuse box or on the steering column.

▶ Remove covers or shrouds as necessary.

▶ Most types of flasher unit simply pull out of the socket.

▶ Replace by pushing the new unit into the holder.

▶ Make sure the new unit is the correct one for the vehicle. Note that more powerful units may be required if a towing socket is fitted.

▶ After renewal, make sure indicators and hazard lights operate.

Job card

Technician/learner name & date	Make and model	VIN no.		Reg. no.	Job/task no.

Customer's instructions/vehicle fault		Mileage		

Work carried out and recommendations (include PPE & special precautions taken)

	Price
Parts and labour	
Total	

Data and specifications used (include the actual figures)

Assessor report

	Assessment outcome	Passed (tick ✓)
1	The learner worked safely and minimised risks to themselves and others	
2	The learner correctly selected and used appropriate technical information	
3	The learner correctly selected and used appropriate tools and equipment	
4	The learner correctly carried out the task required using suitable methods and testing procedures	
5	The learner correctly recorded information and made suitable recommendations	

	Tick	Written feedback (with reference to assessment criteria) must be given when a learner is referred
Pass: I confirm that the learner's work was to an acceptable standard and met the assessment criteria of the unit		
Refer: The work carried out did not achieve the standards specified by the assessment criteria		

Assessor name (print)	Assessor PIN/ref.	Date

Section below only to be completed by the learner once the assessor decision has been made and feedback given

I confirm that the work carried out was my own, and that I received feedback from the Assessor	Learner name (print)	Learner signature	Date

Worksheet 70: Remove and refit windscreen wiper motor

Procedure

Note: this is a generic procedure for a motor that can be accessed from the engine compartment; refer to the specific manufacturer's instructions. Switch off the ignition. Mark the position of the wiper blades with masking tape, and remove the wiper arms.

▶ Raise the bonnet and remove rubber strip and/or covers from the heating/ventilation system area.

▶ Remove wiper motor cover panels.

▶ Remove retaining screws as appropriate and remove the wiring harness plug from the motor.

▶ Unscrew the large nut on the wiper spindles.

▶ Slacken and remove the motor mounting bracket screws.

▶ Manoeuvre the motor and drive linkage out from its fittings, and remove from the vehicle.

▶ Undo the nut on the wiper spindle after marking the position of the crank arm. Unscrew the motor fixing bolts and remove the motor.

▶ Refitting is a reversal of the removal process. However, note the following points.

▶ Connect the motor to the harness and run it (without the linkage) until it stops in the 'park' position as normal. Disconnect from the wiring.

▶ Refit the crank and linkage exactly as it was removed.

▶ After refitting the motor and linkage, run the motor and make sure the movement is correct *before* refitting the arms and blades.

▶ Finally, fit the arms and blades, wet the screen and check for correct operation at all speeds and settings. Check that the blades park correctly.

Job card

Technician/learner name & date	Make and model	VIN no.		Reg. no.	Job/task no.

Customer's instructions/vehicle fault		Mileage			

Work carried out and recommendations (include PPE & special precautions taken)

Parts and labour	Price
Total	

Data and specifications used (include the actual figures)

Assessor report

	Assessment outcome	Passed (tick ✓)
1	The learner worked safely and minimised risks to themselves and others	
2	The learner correctly selected and used appropriate technical information	
3	The learner correctly selected and used appropriate tools and equipment	
4	The learner correctly carried out the task required using suitable methods and testing procedures	
5	The learner correctly recorded information and made suitable recommendations	

	Tick	Written feedback (with reference to assessment criteria) must be given when a learner is referred
Pass: I confirm that the learner's work was to an acceptable standard and met the assessment criteria of the unit		
Refer: The work carried out did not achieve the standards specified by the assessment criteria		

Assessor name (print)	Assessor PIN/ref.	Date

Section below only to be completed by the learner once the assessor decision has been made and feedback given			
I confirm that the work carried out was my own, and that I received feedback from the Assessor	Learner name (print)	Learner signature	Date

Worksheet 71: Inspect and measure wiper motor operation

Procedure

Note: wipers should not be operated for long periods on a dry screen. Take extra care when operating wipers with the linkage exposed as it is easy to trap your hands and cause injury.

▶ Run through the washer and wiper operations to check for correct operation. Most wipers have a slow, fast, intermittent and wash/wipe facility.

▶ Check that the blades park when the switch is turned off.

▶ Switch off the ignition.

▶ Connect an ammeter in series with the motor supply. This can be in place of the fuse if necessary, by using suitable adapters.

▶ Run the wipers at each speed after wetting the screen (using the washers is probably easiest).

▶ Measure the current draw. Readings will vary, but figures in the region of 12A or more are to be expected.

▶ A low reading would indicate a high resistance in the circuit.

▶ Inspect the circuit and make sure connections are clean. Connecting a voltmeter across a connection is a good way of testing it. A reading of almost zero volts is to be expected when the motor is operating.

Job card

Technician/learner name & date	Make and model	VIN no.		Reg. no.	Job/task no.

Customer's instructions/vehicle fault		Mileage			

Work carried out and recommendations (include PPE & special precautions taken)

Parts and labour	Price
Total	

Data and specifications used (include the actual figures)

Assessor report

Assessment outcome	Passed (tick ✓)
1	The learner worked safely and minimised risks to themselves and others	
2	The learner correctly selected and used appropriate technical information	
3	The learner correctly selected and used appropriate tools and equipment	
4	The learner correctly carried out the task required using suitable methods and testing procedures	
5	The learner correctly recorded information and made suitable recommendations	

	Tick	Written feedback (with reference to assessment criteria) must be given when a learner is referred
Pass: I confirm that the learner's work was to an acceptable standard and met the assessment criteria of the unit		
Refer: The work carried out did not achieve the standards specified by the assessment criteria		

Assessor name (print)	Assessor PIN/ref.	Date

Section below only to be completed by the learner once the assessor decision has been made and feedback given			
I confirm that the work carried out was my own, and that I received feedback from the Assessor	Learner name (print)	Learner signature	Date

Worksheet 72: Check central door locking and alarm operation

Procedure

Note: different systems operate in different ways so check specific data as necessary.

▶ Use a scan tool where appropriate to check for alarm/central locking stored fault codes.

▶ Close all the doors and operate the central locking from the driver's door lock using the key manually. All doors and the tailgate should lock. If a double locking system is fitted, turning the key again double locks all openings.

▶ Check manually that all doors and openings have locked.

▶ Repeat the above procedure using the remote key if available.

▶ Repeat again from the passenger's door lock.

▶ Open one of the windows and then fully lock the car.

▶ Reach inside the car. If a movement sensor is incorporated, the alarm will sound! If not, reach in and open the door from the inside. The alarm should now sound! Press the remote or use the key in the driver's door to reset.

▶ Close all windows and lock the car.

Job card

Technician/learner name & date	Make and model	VIN no.	Reg. no.	Job/task no.

Customer's instructions/vehicle fault	Mileage	

Work carried out and recommendations (include PPE & special precautions taken)

Parts and labour	Price
Total	

Data and specifications used (include the actual figures)

Assessor report

	Assessment outcome	Passed (tick ✓)
1	The learner worked safely and minimised risks to themselves and others	
2	The learner correctly selected and used appropriate technical information	
3	The learner correctly selected and used appropriate tools and equipment	
4	The learner correctly carried out the task required using suitable methods and testing procedures	
5	The learner correctly recorded information and made suitable recommendations	

	Tick	Written feedback (with reference to assessment criteria) must be given when a learner is referred
Pass: I confirm that the learner's work was to an acceptable standard and met the assessment criteria of the unit		
Refer: The work carried out did not achieve the standards specified by the assessment criteria		

Assessor name (print)	Assessor PIN/ref.	Date

Section below only to be completed by the learner once the assessor decision has been made and feedback given			
I confirm that the work carried out was my own, and that I received feedback from the Assessor	Learner name (print)	Learner signature	Date

Worksheet 73: Remove and refit central door locking actuator

Procedure

Note: this is a generic procedure; refer to the specific manufacturer's data for detailed instructions.

▶ Fit a memory keeper and disconnect the battery earth/ground.

▶ Remove the interior door handles, lock lever covers, window winder and plastic boxes, etc. as required.

▶ Using a flat forked lever under the plastic clips, remove the interior door trim panel.

▶ If fitted, remove the plastic waterproofing cover. Take care not to tear this or be prepared to renew as required.

▶ Make sure the window is fully closed to allow access into the door cavity.

Note: the inside edges of the door structure are often sharp. Wear protective gloves as necessary.

▶ Trace the wires from the lock actuator and disconnect the multiplug. On some cars, this is part of the actuator, but on others, it may be inside the sill area or even under the carpets beneath the seats.

▶ Unscrew the actuator fixings and unhook it from the pull rod.

▶ Remove the actuator from the vehicle.

▶ Refitting is a reversal of the removal process.

Job card

Technician/learner name & date	Make and model	VIN no.	Reg. no.	Job/task no.

Customer's instructions/vehicle fault		Mileage		

Work carried out and recommendations (include PPE & special precautions taken)

Parts and labour	Price
Total	

Data and specifications used (include the actual figures)

Assessor report

Assessment outcome		Passed (tick ✓)
1	The learner worked safely and minimised risks to themselves and others	
2	The learner correctly selected and used appropriate technical information	
3	The learner correctly selected and used appropriate tools and equipment	
4	The learner correctly carried out the task required using suitable methods and testing procedures	
5	The learner correctly recorded information and made suitable recommendations	

	Tick	Written feedback (with reference to assessment criteria) must be given when a learner is referred
Pass: I confirm that the learner's work was to an acceptable standard and met the assessment criteria of the unit		
Refer: The work carried out did not achieve the standards specified by the assessment criteria		

Assessor name (print)	Assessor PIN/ref.	Date

Section below only to be completed by the learner once the assessor decision has been made and feedback given			
I confirm that the work carried out was my own, and that I received feedback from the Assessor	Learner name (print)	Learner signature	Date

Worksheet 74: Remove and refit fuel tank sender unit

Procedure

▶ Fit memory keeper and disconnect the battery earth/ground lead.

▶ Support vehicle on hoist.

▶ Drain fuel from the tank using special equipment.

▶ If necessary, disconnect fuel lines and filler components.

▶ Remove fuel tank (some senders are accessible without removing the tank, check manufacturer's data).

▶ Remove wires from sender unit.

▶ Undo the ring of bolts and remove sender.

▶ Use a new gasket and sealant when refitting.

▶ Ensure the new sender is positioned correctly.

▶ Refitting is a reversal of the removal process.

Job card

Technician/learner name & date	Make and model	VIN no.		Reg. no.	Job/task no.

Customer's instructions/vehicle fault	Mileage	

Work carried out and recommendations (include PPE & special precautions taken)

Parts and labour	Price
Total	

Data and specifications used (include the actual figures)

Assessor report

Assessment outcome		Passed (tick ✓)
1	The learner worked safely and minimised risks to themselves and others	
2	The learner correctly selected and used appropriate technical information	
3	The learner correctly selected and used appropriate tools and equipment	
4	The learner correctly carried out the task required using suitable methods and testing procedures	
5	The learner correctly recorded information and made suitable recommendations	

	Tick	Written feedback (with reference to assessment criteria) must be given when a learner is referred
Pass: I confirm that the learner's work was to an acceptable standard and met the assessment criteria of the unit		
Refer: The work carried out did not achieve the standards specified by the assessment criteria		

Assessor name (print)	Assessor PIN/ref.	Date

Section below only to be completed by the learner once the assessor decision has been made and feedback given			
I confirm that the work carried out was my own, and that I received feedback from the Assessor	Learner name (print)	Learner signature	Date

Worksheet 75: Remove and refit temperature sensor

Procedure

▶ Fit covers as required to keep paintwork clean.

▶ Remove cap from coolant header tank (vehicle must be cold) and then replace it. This will ensure that there is no pressure in the system and will minimize coolant loss.

▶ Disconnect wire(s) from sender unit.

▶ Prepare new unit for installation by applying sealant to the threads (if required). Follow manufacturer's guidance.

▶ Remove old sensor from the engine (usually at the front of the head) and replace new one immediately.

▶ Torque to specified value.

▶ Reconnect wire(s).

▶ Top up cooling system.

▶ Start engine and run up to temperature. Check for correct gauge operation and leaks.

Job card

Technician/learner name & date	Make and model	VIN no.		Reg. no.	Job/task no.
Customer's instructions/vehicle fault		Mileage			

Work carried out and recommendations (include PPE & special precautions taken)

Parts and labour	Price
Total	

Data and specifications used (include the actual figures)

Assessor report

	Assessment outcome	Passed (tick ✓)
1	The learner worked safely and minimised risks to themselves and others	
2	The learner correctly selected and used appropriate technical information	
3	The learner correctly selected and used appropriate tools and equipment	
4	The learner correctly carried out the task required using suitable methods and testing procedures	
5	The learner correctly recorded information and made suitable recommendations	

	Tick	Written feedback (with reference to assessment criteria) must be given when a learner is referred
Pass: I confirm that the learner's work was to an acceptable standard and met the assessment criteria of the unit		
Refer: The work carried out did not achieve the standards specified by the assessment criteria		

Assessor name (print)	Assessor PIN/ref.	Date

Section below only to be completed by the learner once the assessor decision has been made and feedback given			
I confirm that the work carried out was my own, and that I received feedback from the Assessor	Learner name (print)	Learner signature	Date

Worksheet 76: Check operation of heating and ventilation system

Procedure

▶ Start the engine and run until it is warm (use extraction if indoors).

▶ Check that the booster fan runs at all speeds. Switch off AC if fitted.

▶ Set the temperature control to cold and the fan speed to a medium setting.

▶ Run through all direction settings and check that *cool* air is supplied.

▶ Set the temperature control to hot and the fan speed to a medium setting.

▶ Run through all direction settings and check that *hot* air is supplied.

▶ Check that a range of temperatures can be selected and that external or recirculated air can be used.

▶ Make sure all ventilation grills open and allow directional control.

▶ Check heated rear screen operation.

▶ Check heated front screen operation (if fitted).

Job card

Technician/learner name & date	Make and model	VIN no.		Reg. no.	Job/task no.

Customer's instructions/vehicle fault		Mileage			

Work carried out and recommendations (include PPE & special precautions taken)

Parts and labour	Price

Total	

Data and specifications used (include the actual figures)

Assessor report

	Assessment outcome	Passed (tick ✓)
1	The learner worked safely and minimised risks to themselves and others	
2	The learner correctly selected and used appropriate technical information	
3	The learner correctly selected and used appropriate tools and equipment	
4	The learner correctly carried out the task required using suitable methods and testing procedures	
5	The learner correctly recorded information and made suitable recommendations	

	Tick	Written feedback (with reference to assessment criteria) must be given when a learner is referred
Pass: I confirm that the learner's work was to an acceptable standard and met the assessment criteria of the unit		
Refer: The work carried out did not achieve the standards specified by the assessment criteria		

Assessor name (print)	Assessor PIN/ref.	Date

Section below only to be completed by the learner once the assessor decision has been made and feedback given			
I confirm that the work carried out was my own, and that I received feedback from the Assessor	Learner name (print)	Learner signature	Date

Worksheet 77: Inspect heater controls and heater blower motor

Procedure

▶ Obtain information from the driver regarding heater function. Turn the heater control to the coldest setting (with air conditioning off). Check that the engine warm up time is correct by feeling that the top hose remains cool until the thermostat begins to open. The top hose should then rapidly heat up. If in doubt, apply a thermometer to the top hose or in the radiator header tank to check engine temperature.

▶ Check engine heat temperature range. Leave on the cold setting and run the blower motor at all speeds in its range. Does air at the outside temperature flow from the heater ducts? If yes, the cold adjustment is correct.

▶ If warm air flows from the heater, check and adjust the temperature control cable to the air mixture flap in the heater or to the water control valve. If this does not give cold air, the seals on the control flap or in the water control valve are probably defective. Turn the heat control to the hottest position and check for hot air from the heater ducts with the blower motor operating.

▶ Check that as the heat control is moved slowly towards the cold position, the air temperature from the heater ducts becomes steadily cooler. For water control valve and automatic systems some delay in reducing the temperature may occur. To check automatic temperature systems allow the engine to cool. Set the temperature control to an intermediate position and check that the air temperature remains stable as the engine warms up.

▶ Check the air directional control by selecting each position in turn and feeling for air flow from the appropriate positions. If incorrect, move the control slightly to either side of the proper position to find the adjustment error and then adjust the cable accordingly.

▶ If the heater blower motor fails to operate, check the fuse and electrical feed to the fuse box. If correct, check for voltage at the motor terminal block with a voltmeter or test lamp. Follow a wiring diagram for feed and earth or ground cables. Check earth or ground continuity.

▶ If voltage is correct at the terminal block and earth or ground is continuous, check the motor brushes and commutator for condition. Brushes may be available as a replacement part but usually a new or replacement motor will be required.

Job card

Technician/learner name & date	Make and model	VIN no.		Reg. no.	Job/task no.
Customer's instructions/vehicle fault		Mileage			
Work carried out and recommendations (include PPE & special precautions taken)					
Parts and labour					Price
Total					
Data and specifications used (include the actual figures)					

Assessor report

	Assessment outcome	Passed (tick ✓)
1	The learner worked safely and minimised risks to themselves and others	
2	The learner correctly selected and used appropriate technical information	
3	The learner correctly selected and used appropriate tools and equipment	
4	The learner correctly carried out the task required using suitable methods and testing procedures	
5	The learner correctly recorded information and made suitable recommendations	

	Tick	Written feedback (with reference to assessment criteria) must be given when a learner is referred
Pass: I confirm that the learner's work was to an acceptable standard and met the assessment criteria of the unit		
Refer: The work carried out did not achieve the standards specified by the assessment criteria		

Assessor name (print)	Assessor PIN/ref.	Date

Section below only to be completed by the learner once the assessor decision has been made and feedback given			
I confirm that the work carried out was my own, and that I received feedback from the Assessor	Learner name (print)	Learner signature	Date

Worksheet 78: Check operation of in-car heating including air distribution and fan operation

Procedure

▶ Set heat control to the coldest setting and run the engine up to normal operating temperature. As the engine heats up, check the air distribution control and heater blower motor operation.

▶ Check inside the engine compartment that the heater feed and return hoses are hot and cold, respectively, for water valve types and that both hoses are hot for air mixing types.

▶ Keeping the heat setting on the coldest position, turn on the blower fan and check that the airflow from the heater is similar to the outside air temperature.

▶ Check that the fan increases in speed for each of the speed settings and that the airflow increases as the motor speed increases.

▶ Move the heat setting progressively from the coldest to the hottest setting and feel the airflow for a gradual increase in temperature. Do not operate the air conditioning system if fitted.

▶ Check the air distribution control by moving to each position in turn and checking that airflow is directed from the correct outlets.

▶ For air-cooled engines it is important to check for exhaust gas leakage inside the heat exchanger. Check the heater outlets for odour.

▶ Alternatively, for increased safety, use a gas analyser with the probe inserted into a side window. Seal the window opening with tape. Set the heater controls to the highest temperature. Run the engine and look for the presence of exhaust gases in the vehicle interior. Check for CO, which is toxic.

Job card

Technician/learner name & date	Make and model	VIN no.		Reg. no.	Job/task no.

Customer's instructions/vehicle fault	Mileage				

Work carried out and recommendations (include PPE & special precautions taken)

	Price
Parts and labour	
Total	

Data and specifications used (include the actual figures)

Assessor report

	Assessment outcome	Passed (tick ✓)
1	The learner worked safely and minimised risks to themselves and others	
2	The learner correctly selected and used appropriate technical information	
3	The learner correctly selected and used appropriate tools and equipment	
4	The learner correctly carried out the task required using suitable methods and testing procedures	
5	The learner correctly recorded information and made suitable recommendations	

	Tick	Written feedback (with reference to assessment criteria) must be given when a learner is referred
Pass: I confirm that the learner's work was to an acceptable standard and met the assessment criteria of the unit		
Refer: The work carried out did not achieve the standards specified by the assessment criteria		

Assessor name (print)	Assessor PIN/ref.	Date

Section below only to be completed by the learner once the assessor decision has been made and feedback given			
I confirm that the work carried out was my own, and that I received feedback from the Assessor	Learner name (print)	Learner signature	Date

Printed in the United States
by Baker & Taylor Publisher Services